The Man Who Loved
SPEARFISHING

STAN CATES

ISBN: 978-1-66785-587-5

CHAPTER 1

The Beginning

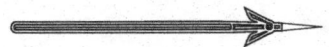

Returning to Corpus Christi Texas in 1961 after 2 years as a traveling electrician working on missile sites in Cheyenne Wyoming. and Topeka Kansas, I saw an ad in the Caller Times about scuba classes. I signed up. The classes were held at the local YMCA. The instructor was Jim Copeland. I was an avid fisherman and hunter and had seen some films on spearfishing. My only thought from the beginning was to get in the ocean and shoot a fish. Early on, Cousteau's men were avid spear fishermen.

The classes were very good and stressed safety. We did a lot of snorkeling besides buddy breathing and towing an unconscious diver. At the end we played capture the flag with the lights off. Weights on both ends of the pool were to be retrieved with the opposing team tearing off your mask to stop you. You had to return to your end of the pool to retrieve your mask. Great fun. We had a mix of the new single hose regulators and the Cousteau 2 hose units. I chose the single hose sportsways regulator. We all thought we were brave, and might be eaten by sharks when we dove the Gulf of Mexico. Jim's brother Rick and a future dive buddy Bob Turner were in this class.

Jim Copeland's YMCA scuba class 1961

Jim could tell I was raring to go diving and learn what spearfishing was all about. He set up a date and we headed out to Padre Island. The technique was to back the boat up to the surf and hit the brakes and start the boat off into the shallow surf and quickly pull forward. After loading the boat with diving gear, you walk it out to a gut and start the engine. Then watch the surf break and try to time take off to coincide with diminished waves (good luck with that). We went out about 3 miles to the cluster (a group of about 7 rigs) The water was rough and dirty brown and I was getting sea sick. I swam a tie up rope over to the rig. We put our gear on and fell in the water. Jim with his gun and me as an observer. My god this is horrible I thought to myself as I fought to keep from puking. The visibility was about 5 ft.,Jim swam down looking for jewfish (now called goliath grouper) and returned empty handed. I'm thinking this ain't for me as we headed home. At that time Jim was selling dive gear out of his closet, he did not have a dive shop yet.

A couple of weeks later Jim set up a trip out of Port Aransa on a larger boat with more people. We headed out, and as we crossed over the jetty extensions I could see the bottom, much better day. We went to the double platforms and the visibility was at least 75 ft. It was beautiful with multitudes of fish in sight, we were in mother oceans aquarium, I was captured for life.

I BECOME THE INSTRUCTOR.

I took my best friend Jimmy Heidland to a pool and taught him to scuba dive. We bought used spearguns. Jimmy had a small boat,motor and trailer and we headed to Padre Island. The visibility was fair and a cobia swam by me and I shot it, it took off around a piling immediately cutting my line on the barnacles. I was sick and disgusted, my first fish gone in a flash. Back to the drawing board and learn how to use ss cable. I was always in too big a hurry. We had fishing gear and there were so many king mackerel circling the rig you couldn't even let your bait dangle over the water, they were ravenous and would come out of the water after our bait. We were clueless about offshore fishing and quickly lost our fish and some tackle as they stripped our lines. Wow, these ain't trout I thought. On our next trip I got my cobia (we all called ling in those days) . I had a good shot but was afraid of losing another fish so I followed as best I could trying not to put too much pressure on it. In the end at the boat my mask was full of blood from the up and down pressure changes on my sinuses. I was very very happy. We then started teaching some of our other friends to scuba dive. It's pretty simple, breathe and kick your feet. If a person wasn't a natural waterman, we didn't mess with them. Snorkeling is mandatory, of course we stressed the danger of embolism and everyone got charts with no decompression bottom times. The rigs we dove were only 60 ft deep. Over time, none of our friends drowned but some managed to turn their boats over in the surf and lose diving gear.

First fish, a nice cobia

Somehow we acquired a high school kid as a 3rd dive buddy. Steve Shaw had been certified by Copeland. He had no money to help with expenses but we liked his aggressive attitude and he helped us rig up our guns for jewfish with larger SS cable, a slip tip and rope for a riding rig. Once you shoot, the gun is free, then you stick an arm through the rubbers and hang on to the rope. Over time, on other trips with Copeland I watched other guys land two large jewfish and got braver. They looked like Volkswagens to me. They told me to hang on, keep your face into the current so your mask doesn't wash off until it slows down. Then grab it in the eyes and swim it up. Sounded simple enough.

FIRST JEWFISH.

On another trip to the rigs I spotted a smaller jewfish I thought I might handle. I approached it and placed a shot in the side of it's head and it took off. I stayed with it till it slowed down. Somehow I ended up in front of it, my adrenaline pumping wildly I grabbed it in the eyes GULP, it gulped my arm up to the elbow. I panicked and jerked my arm back, losing a bunch of skin. Their lips are coated with tiny spines. I was a bloody mess. Still pumped, I quickly swam around behind it and grabbed its eyes again, it shook it's head violently while I struggled towards the surface. For the rest of my life whenever I shot a fish and didn't stone it, I went into the fight mode (do or die) This beautiful fish weighed around 250 lbs and provided lovely filets. jewfish are excellent food.

First jewfish

I LEARN NOT TO HUG COBIA

We had a lot of cobia in our rigs and we often shot them in poor visibility. I didn't want them out of sight once I shot one (dumb, they calm down when allowed to swim down a little). I had a long sleeve shirt on (called rash guards today) when I shot my next cobia. I pulled it to me and as it struggled, hugged it to my body to subdue it. Big mistake. They have a row of small sharp dorsal spines and as they spin and struggle they will leave several small slices in your skin.

JIMMY HEIDLAND GETS MARRIED.

Jimmy married Agnes who then informed me she was taking over Jim and his life (and she did) I lost my boat ride and dive buddy. Jim and I had gotten more and more experience over the previous 3 years and routinely landed 2 and 300 pound jewfish, lots of cobia and I had taken 3 or 4 cubera. We had also met some college kids who were avid spearo's, one of them was Billy Causey and his wife Glenda. A local PI attorney Guy Allison was the envy of all of us. He had money and a large fast boat. Guy used compressed air guns and a parachute strap for a riding rig. The more we dove the more we got used to brown water and shooting in low visibility. When agitated, jewfish convulse their gills and emit loud booms. Once I was near the bottom with 2 ft of visibility and heard the booming. As I inched along a horizontal it got louder. As I peered forward I saw a patch of yellow. Straining to see, I made out its eyes. I was about to stick my head in a jewfish mouth. To hell with this, I gave up. JF eat anything. They have grinders in their throat they use to crush crabs. Years later I had one vomit up a 3 ft shark on the deck of my second boat, they simply inhale whatever they can get near.

Jimmy Heidland the best person I ever knew

FIRST BOAT.

I bought an 18ft Hurricane boat motor and trailer for $800. Now I needed diving buddies. I went through several dive buddies over the next years, but never found one like Jimmy Heidland. One of the guys I teamed up with was a liar, a coward and a thief. Copeland had opened a dive shop in Six points and I stopped by often. The word was out that there was a monster jewfish at the Fox rig and people were afraid to shoot it. I called Mark and we took off for the beach near the Fox rig. I had rigged a Scubapro tube gun with 4 bands and hoped some day to stone a jewfish. (never happened) We got in the water and saw the monster. (I later estimated at 800 lbs or more) Fox is a small tripod rig. This fish filled it up. I didn't have a head shot. I told Mark I would squeeze between 2 pilings for a brain shot, back out, swim around and into the stunned fish and grab my rope. Meanwhile he could shoot it from the side and we would ride and subdue it. Brilliant plan.

Squeezing in and looking down I pointed at the stone zone and squeezed my trigger as hard as I could. These guns were not designed for 4 bands. The gun finally fired. I backed out and started to swim into the rig. The monster was coming out. It ran over me and knocked me against a piling, my cable pressed against my chest, in a panic I pushed the cable down away from my neck, fearing it would pinch my head off. (this is a bad memory) As my cable then rope reached my feet the fish was gone. I thought, idiot, if you had grabbed the rope you would have had it in open water. However the rope was moving so fast I really could not have held on. Mark showed up dazed and confused and had lost his Rolex. We found his watch and headed home. We stopped by a bait stand and when we left Mark had a new pair of sunglasses. I said "I didn't see you pay for those" He grinned at me.

Later at my house I laid out 5 $20's on my seat to dry out. My wife and kids were out of town. We showered and cleaned up. After he left I noticed a $20 missing. He swore he didn't take it. He was trying to get in the union and I was on the executive board. I told him he would never get in the union. He finally made a weak excuse about mixing our stuff up and gave it back. That's a liar and a thief. Coward. On an earlier trip when I had a smaller gun I came up with another brilliant plan on how to shoot a jewfish. I would shoot it in the eye and into its brain and stone it. We came across one in the 200 lb range. I worked around for the best angle and shot it in the eye. The complete shaft came out of its gills and the fight was on. As I slowed it down I was running out of air, I tried to hand my gun to Mark, he pulled away in fear as if I was trying to harm him. I quickly tied my cable around an anode and streaked to the surface to change tanks. When I came back the fish had been resting and the fight started all over again. When we got the fish into the boat, we saw it had been hit in the back with a shotgun and had a huge hole full of rotten meat. I cut the rotten meat out and the rest was fine food. This made me feel better about jewfish I had previously lost. They are sturdy creatures and if not mortally wounded they heal. A funny and final note on Mark. While hanging out near Bob Hall pier one day Mark was there surfing. I was watching him when he fell and jumped up quickly, the board hit him in the head and he went down in the brown water. I kept my eyes on the exact spot and waded there as quickly as I could. I felt down and around in the water and found him. I picked him up by the armpits and started dragging him ashore. People just stared at me as I yelled "I could use some help"

THE COBIA THAT KILLED OUR KITTEN

I came home from diving one day in the midday heat. My wife Jackie was at her wits end with 3 young boys. I was hot and exhausted. I asked her if she would clean my fish, big mistake. You shoot em, you clean em she yelled at me. It was so scorching hot I decided to bring the fish into the kitchen sink in the air conditioned house. It was about a 40 pounder. I had it on the sink, it was very slippery. Our twin boys were watching me along with their new kitten. The fish slipped off of the sink and fell directly on top of the curious kitten, smashing it. Its lungs were coming out of its mouth. The boys started screaming, Jackie came in screaming, everyone was screaming at me. I quickly took the cat outside, it was still alive, I had to finish it with a 22. What a mess. I finished cleaning the fish outside.

My oldest son Stan, was a big help cleaning the boat. He likes to remind me of the time he was helping and stepped into a gob of jewfish slime and had a bad fall. We got jewfish in the boat by putting a rope in their lip and lifting, as they came up, the boat would tip down until we reached the balance point. Then we would get out of the way, with the last pull they would slime in and crash on the deck leaving slime everywhere.

I RETRIEVE GUY ALLISON'S FISH.

A few of us were invited to go diving with Guy on his cool boat. It was a clear water day and I was doing more snorkeling than scuba. Guy came back with a mashed finger from fighting a jewfish. He said he needed someone to retrieve it. I had the most air left in my tank so I was selected. Guy led me down to the spot on the bottom where he said the fish was. His cable disappeared into solid murk. I felt my way down into zero visibility. I could feel the fish and the tangled cable mess. I tried and tried to unravel the mess. Finally I collected the cable into a ball and started passing it through tangles and knots and was able to free it. Guy kept hold of my shoulder the whole time. There was loud cheering from the boat as we brought the fish up. None of us were offered a filet. Guy was a tort lawyer and made millions. He wouldn't let us pay a dime to go with him. But when called on to work on his boat, you better show up.

NOGI 1963

I saw a brochures advertising NOGI (New Orleans Grand Isle) International Spearfishing. Pictures of seasoned divers posing with large fish intrigued and excited me. I had to go. I pleaded and pleaded with Copland and finally talked him into going. We had never been to the French Quarter and those were my drinking days. You couldn't watch a show without buying a drink. Jim is very religious. He decided to buy a drink and I would drink both. This didn't last long till he said he was leaving. We agreed to meet up later and I kept going from joint to joint buying drinks until I had enough. On the way down to Grand Isle I had a flat. Jim had to use a towel to wave mosquitoes away while I changed the tire.

The first days diving exhausted us. We had to learn how to drink plenty of water, then we were ok for the last 2 days. We were used to quick, leave at daylight trips off the beach and back in by 9am. I had never dived water deeper than 60 ft. At a clear water rig I was down 100 ft and was fine. An amberjack came by and I shot it with my Sportsways gun. (the Bandido today and beefed up) The handle snapped off and the fish was swimming down and away with my spear and barrel. I took off chasing madly after the rest of my gun. I caught it just as it was about to disappear over a horizontal. Suddenly I realized I was smothering. This regulator was starving me for air. It was like you ran to the top of a hill and you were trying to catch your breath through a straw. I headed for the surface swimming as hard as I could holding my fish and gun. As I rose I exhaled a little to keep from embolizing. I don't know how I made it. I lay on the surface taking deep breaths for quite a while.

On the last day we hadn't shot a NOGI trophy fish so I decided to shoot a jack crevalle. I was outside the rig fighting it when the sharks showed up. Every time they circled around they got closer. Jim was with me, but when I looked around he was gone. Finally I wrapped my cable around an anode and headed back to the boat. As I approached the last piling there was Jim below me peeking around both sides of the piling for sharks. I swam down behind him and grabbed his leg, he whirled around in panic, his eyes as big as saucers. It was hysterically funny. Back on the boat I talked some experienced divers into retrieving my gun, handing them some cable cutters. They said they would bring it all back, but accepted the cable cutters. They returned with only the gun and said those sharks were getting aggressive. I never became close friends with Jim. We were always friendly, but he was and is always a businessman. Jim started out very poor and building and running a dive shop from scratch is very hard work. To keep from starving in the winter, Jim got into the ski business. Promotion became his full time activity. My oldest Son Stan, worked for him a short time in later years.

JAY RIFFE COMES TO TOWN.

Stopping by Copeland's dive shop one day in six Points I met Jay Riffe. Jay was driving a portable machine shop in an 18 wheeler. He had already started making spearguns. We hit it off and I invited him to go honky tonkin with me. My wife and kids were out of town, so we met up later to visit my favorite dance bar. Jay danced with some of the local gals and we drank some beer. The next day he showed up at my work place (Reynolds Metals) to demonstrate machine tools. I had no clue about real free divers. In fact

I thought freedivers shooting real fish in the oil rigs would be a daunting task. Jay told me about his brother John who was making a reel for spearguns. I ordered one. After displaying his wares in the Corpus area Jay was off to Houston where he met up with Mac Blaker. I believe they went free diving. Blaker had a dive shop and at the time was known for spearfishing the Flower Garden reefs. He had a customer whose father owned a fleet of oil platform service boats. Later Blaker became known for his surf boards and surfing events in Galveston. Last I heard he had moved to Hawaii.

JIMMY AND LUKE, BILLY AND GLENDA.

My nemesis was Jimmy Clark. He and Billy Causey built a neat high bow boat for surf launching. I had to beat them to the rigs or they would get the best fish. Billy and Bob Turner had gone to work for the Padre Island Seashore. After finally running out of diving buddies I called Jimmy and acquired a life long friend. We made many trips in his and my boats. I made a winter trip down the beach to 7 ½ fathom reef with Billy and Glenda. Billy was working on a paper for his degree about the fish on this reef. Billy drug his boat for surf launching. It's a long drive. When we got to the reef (about 2 miles offshore) Billy ran his transic line counting fish and I of course started shooting snapper. This reef had ragged tooth sand sharks but they were not aggressive. I had acquired about 3 snapper when my wetsuit came unsnapped at the waist, letting cold water in. Soon I started cramping from the cold. I started heading for the boat, then both legs cramped and doubled up, my knees up towards my chest and I ran out of air. I rolled over on my back, spit my regulator out and using my snorkel started undulating towards the boat, bobbing up and down with my fish and gun intact. Billy helped me in the boat. I never considered drowning. Maybe this is what they mean when calling a person a "waterman". Panicking divers can not only die, but drown those trying to save them. I was growing very comfortable in mother ocean.

Jimmy Clark was a welder. He had developed the best jewfish riding rig I have ever seen. We used SS shafts. They would bend, but not snap like the brittle spring steel shafts. Jimmy would weld a small ss loop to the shaft near the tip. You then passed your cable through the loop, next came a swedge clamped to the cable,12 to 16 inches later your slip tip. The cable was about 10ft long with the slip tip on one end and the rope on the other. When you shoot a jewfish you jerk back hard, the swedge pulls the shaft out because it can't pass back through the little ss loop, leaving the slip tip in the fish sideways and the shaft hanging on the cable. I was later told that the weakest link in this rig was my armpits.

VERACRUZ

Billy and Glenda, Jimmy and Luke made annual 2 week trips to a small lighthouse island (Isla de en Media) just below Veracruz and out in front of Punta Antone Lizardo. They would drag the boat down from Corpus with all their gear and a small compressor. We all agreed I would fly down and spend a week with them. Youth and energy are wonderful. Southwest Airlines was Trans Texas in those days (around 1967) I placed my spearguns behind the pilots seat. They examined them and were jealous of where I was going. It was a great trip with beautiful reef and lots of fish. I mostly snorkeled and shot small snapper. Glenda would excitedly point out fish I was trying to sneak up on for a shot. Their friend Don Reese showed up for a day. Don later opened a dive shop in Fort Worth. Billy even managed to land a jewfish. We scuba dived through deep beautiful canyons. Thousands of red and black crabs would come out at night. I saw huge cubera snapper and was happy not to get a shot with my small gun.

The lighthouse keeper Pancho, would have limes, salt and tequila shots waiting for us every day when we came in from diving. He nicknamed me Largo, I was a skinny 6ft 2 inches. He called Glenda, Linda. I agreed, she got so brown down there, strangers assumed she was mexican.

Glenda and Luke were eyeballing my plane ticket as I prepared to fly home. They had a grueling trip ahead pulling the boat all the way back to Corpus. Billy made Glenda ride in the back of the truck to keep me company on the way to the airport. I was falling in love with Glenda.

I SAVE LUKE

I was going spearing one day and Jimmy couldn't go but Luke wanted to go. I picked up another of Billy Causey's college friends and we took off for the cluster. The surf was up pretty good that day. I always had toilet paper on hand for beach launchings. Everyone got used to waiting for this grand event. I had developed a nervous condition where I had to have a bowel movement before launch. Luke and Mike waited while I ran behind the dunes to do my thing.

We took off and just into the 3rd sand bar waves my old evinrude died. I yelled jump as I bailed out. Luke couldn't swim. She could snorkel and scuba but never learned to swim. I had the bow rope in my hand to keep the pointy end facing the waves as we backed to the beach. The heaving boat came down on Luke's head and stunned her. She appeared in serious trouble. I didn't want to let go of my boat, fearing it would flip over, but I was able to grab Luke with my left hand and hang on to her. As we backed up my feet touched the bottom a few times and I pushed up. Back in the first gut we reassessed. I got the motor started, seemed to be running ok and we took off again. Spearfishing is serious fun and pushes idiots to extremes. Luke loved to tell the story about when I had to choose between saving my boat or her. I got a nice 200 pound JF that day. We always shared steaks.

CHAPTER 2

Freeport Texas

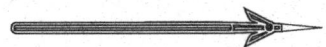

I had been told in high school I had an exceptional IQ, but I needed to apply it. Failing my physical for Annapolis because of a crooked frozen little finger and bad teeth with gaps I couldn't afford to fix, I went into the electrical trade. After apprentice school I signed up for a correspondence course from DeVry Tech. It was excellent with lots of hands on bread board work. I then built a vacuum tube volt ohm meter and an oscilloscope. On my last lesson I used my first solid state device. Times were a changin. As a journeyman I was considered the best control man in the union. I noticed some of the top engineers at Reynolds Metals were GI's with an ICS diploma in Engineering. It was recognized by most industry but not states or colleges. Because of my 4 years apprenticeship and DeVry Tech diploma ICS gave me credit for 2 years and I signed up with their EE power option program. ICS was a torture test. DeVry had been graded by computer. ICS required all work and calculations shown. Someone had to sit and grade each lesson. This made it expensive and very time consuming. I still have my ICS lesson on transformers. All lessons were developed by college professors. with PE licenses. I completed my engineering studies and received my diploma in 1969.

I married my first wife Jackie Austin while she was a junior in high school. She was 17 and I was 21. Her mother gave her to me to raise, it was a case of dumb and dumber. She was beautiful with long blond hair. By

1969 we had 3 boys (1 set of twin boys) Her whole family were chain smokers. She was the gorgeous blond with the cigarette which was typical for the 50's. Jackie was an indoors girl. She hated boats, beaches or anywhere that wasn't air conditioned. A friend did a caricature painting of Jackie. It showed her favorite ashtray with a cigarette and her comb. When she combed her beautiful honey blond hair she got attention.

I went to work for a consulting firm in Corpus, Bath and Associates in 69 right after hurricane Celia had destroyed Portland Tx (across the bay from Corpus) . They had a project in Freeport Tx for Dow Chemical on some automated cranes. The control system selected did not account for the atmosphere of chlorine gases and magnesium, it was terribly designed. This became my project and I was traveling back and forth from Corpus to Freeport. Thad Brown was an engineer that had quit Dow to sell B&A engineering to Dow and others. The bottom fell out of sales and Thad was left with no buyers. He borrowed $50K from Bath and started an electrical testing company that became BES, Bath Electrical Systems. I like field work and decided to join Thad in the new company. Testing sounded like fun to me. We moved to Freeport Tx.

I had sold my old Hurricane boat a few years earlier when I took a foreman's job in Chicago. So I was boatless. I bought my next boat (a real POS) a 23 ft Sterncraft with 2 Mercruiser inboard outboards. I was a happy idiot. The Pequod II. I joined the local dive club to meet some dive buddies. Jim Rio and Cliff Woodiel became lifelong friends. Jim was 20 years younger than me and was always ready to go spearfishing. Once I had dive buddies and a boat I lost interest in the dive club, I was never much for meeting to talk about diving. In those days there were lots of rigs out of Freeport. By now I was the old pro with plenty of fish under my belt. We made many trips in the next months. I showed Jim how to rig for jewfish, we always had fish in the freezer. On one trip to the Buccaneer field Jim had shot a nice warsaw and was trying to pull it up. I swam off to look for a fish, not finding anything. I was coming back and Jim was still trying to pull his fish up. I swam down to it, grabbed it in the eyes with one hand, swam up and handed it to him. BTW,when you need both hands to grab their eyes, you know you have a large fish.

THE VA FOGG AND RIG 538

Sitting in a barber shop in old Freeport on a cool day with doors open, we all heard a loud boom. Later we found out a vessel had blown up 40 miles offshore, killing all on board.

Jim Rio and I spotted a rig on the charts that appeared to be almost beyond reach, rig 538, 40 miles South of Freeport. A friend Ronnie Broadus rode along with us. We had found the paradise rig. We couldn't even get to the 100 ft. bottom with warsaw swimming up to meet us. Ronnie, a fisherman, was freaking out as we brought fish after fish to the boat. It was winter when the warsaw are more plentiful. Finally we headed in. I knew we were running low on fuel so I moved in closer to the beach. Sure enough we ran out about 5 miles short of Freeport. We jumped into the surf and I stood and held the boat while Jim hitch hiked for gas. Real spearo's will not be dissuaded by simply not having enough fuel.

On a later trip to 538 Jim and I always had a contest to see who could get the biggest snapper. On this day I could not find one. But there was a beauty in a fish trap that the rig guys had lowered to the bottom. I shot it, then carefully opened the door and threaded it out, but I was caught in the act. I became known as fish trap Cates. This rig is productive today with a school of gray snapper that has lived there and grown for over 40 years. Many years later, in my freediving days I took Guy Nesbitt there when we weren't getting anything elsewhere. We all got gray snapper, some as large as 12 pounds.

DENNIS FINGERS DRY CLOTHES

I kept the Pequod 2 sterncraft down the coast for a few years in Matagorda. There were a lot of rigs down that way. I would recruit a few divers from the Lake Jackson dive club to go. One regular was Dennis Finger. We had to go out of the mouth of the Colorado River. It was tricky because there were sandbars at the opening. Coming in at low tide we often had to jump out and push the boat over a sandbar. Dennis had a bad habit of putting on his dry clothes before we were safely back in the river. I would holler jump and

Dennis would whine "I have my dry clothes on" I also had a habit. Those who hesitate getting in at a rig might not get a fish, so I learned to pee while leaning back against the side of the boat while loading my gun. (well, it's my boat) I teased "while you guys are peeing, I'm in the water"

One day we made one last stop on the way in. I leaned back to load my gun and started peeing, Cliff Woodiel yelled"your peeing on Dennis's dry clothes" I looked down and sure enough, Dennise's dry clothes lay neatly folded just under my pee stream.

THE SAGA OF THE MIDDLEGROUND

Louis Shaeffer, Thomas McDonald and Jim Tatum.

Cliff Woodiel came to me all excited one day. He had met Louis Shaeffer. Louis had a 50 ft steel boat the Aqua Safari and was looking for divers to help salvage the Middleground, a 150 ft. service boat that had sunk near rig A1A out of Port O'Conner. Sounded like fun to me. It was decompression diving in about 175 ft of water. We had those old scuba pro ADC mechanical meters containing a bourdon tube to move the pointer. We thought they were great.

Louis picked us (Jim Rio, Cliff Woogiel and I) up in a van and we were off to adventure. To us this was luxury diving. A cook, bunks, showers and we could carry our spearguns and shoot a fish when we completed a task. We had deco stops at 20 and 10 ft. Louis had a compressor on board to pump air down to the service boat storage tanks and try to lift it. Warsaw and red snapper were on the bottom, with AJ's halfway down and cuda at the deco stops. The warsaw would follow us around as we installed fittings, air hoses and valves to turn the air on and off. I ended up making 18 dives here over the next 2 years. On one dive I had a monster warsaw following me with a buddy half its size. When I got ready to shoot I chickened out and shot the small one (about 150 lbs) directly above and between the eyes. Stoned it. You did not want to be fighting a huge fish while deco diving. On the way up the fish starts swelling and floating as gasses in the fish expand.

I let it float to the surface while holding the handle of my speargun as I stayed at 20 ft. Guys aboard gaffed it and pulled it aboard.

Turns out the Aqua Safari was a charter diving boat running trips to the Flower Garden, Stetson Banks and the VA Fogg. Our reward for working on the Middleground was helping Louis run trips, fill tanks and keep track of errant divers. Don Reese (Fort Worth) Mike Price (Houston, having bought Mac Blakers dive shop after Mac left for Hawaii) and Jim Copeland (Corpus Christi) all chartered trips. Most trips were to the VA Fogg, the ship that had blown up killing all aboard. I met Thomas McDonald, a friend of Louis and we became lifelong frenemies. Tom had made the first dives on the Fogg with Channel 13 News. He had photos of bodies and the destruction. Tom recovered a Rolex laying loose and was able to return it to the widow. My job was to tie up when we arrived. Louis would watch the old loran screen and tell me when to dive. My gun in one hand and rope in the other. I would lean over and fall off the platform head first and swim straight down 100 ft to the bottom, look for the Fogg and tie up. Then I had first shot. In those days the Fogg was full of warsaw and red snapper. We called the large warsaw silver sides. As they grew and aged, they would turn white on their sides. The first few times I saw Louis dive, he came back with silver sides. He had shown me where to hit them for a stone shot.

On one trip to the Fogg it was late in the day and I hadn't dived since tie up when I had chosen a 20+ lb red snapper. I rigged up and took off by myself. (not unusual) It was beautiful on the bottom. There was a shark laying on the bottom, I touched it and it shot away at high speed. It always amazed how sharks could accelerate after cruising slowly around. The Fogg loomed to my left and from out of the gloom to my right swam the biggest Warsaw I had ever seen. I had my largest gun and with 3 bands the trigger pull wasn't too hard. As it swam slowly towards me I floated up above it and placed a shot in the stone zone. I killed it. My adrenalin was flowing and my heart was pumping like crazy. I couldn't believe I had just killed this monster. When I reached the dive platform I had to yell to get some attention, Mike Price and his crew freaked out at the size of this Warsaw. Getting in late and cleaning up I was exhausted. We got so many fish in those days, often cleaning gear and fish till midnight, many fish were taken without pictures.

ALMOST BENT

On one trip to the Middleground Louis promised us we would stop by the rig AIA for some spearfishing at day's end. My deco meter was clear and I was good to go. Jim Rio and I headed down. Jim had to stop at a horizontal, his ears were not equalizing. I headed to the bottom alone. I ran into a 100 lb+ Warsaw and placed a shot near the stone zone. The fish was dazed but not stoned. It swam away shedding my spear, I couldn't believe it. I settled into the mud on my knees to reload, after my 2nd rubber a huge red snapper swam by, I took a quick shot and gathered it up and headed up. I ran into Jim still unable to clear his ears. I handed him my gun and fish and took his loaded gun and headed back to the bottom. Nothing where I had been, I swam towards the rear where multiple pilings were. There were 3 warsaw there. I could see the wound on the one I had hit earlier. I aimed at one and squeezed off a shot WHUMP, the shaft never left the gun. Jim had gotten his short cable to the slip tip wrapped around the muzzle somehow and I had a useless gun with a bent shaft. Two fish swam away, I decided to grab the groggy one I had shot earlier and thought I would wrestle it back to the boat. (those of you who have been narced know what's going on here) after a brief wrestling match a light went off in my brain "you've been down here too long" Back at the dive platform the seas had picked up and Jim said he almost lost my snapper getting aboard. I told him I would have killed him. After a while on board I felt the needles in my left shoulder, but they didn't last long. That's as close to bent as I ever want to get. On a later trip Jim was bent and had to be airlifted to a chamber. Sometimes in heavy seas the only way to get on the dive platform was to time it so you swung your butt over it just when it reached max low. Thrilling times for idiots. Wife and kids at home, no pay and no insurance. Just fun and adventure.

Jim Tatum was a criminal attorney in Houston and a friend of Louis Shaeffer. Louis paired Tatum and I up on dive trips and we became friends. He kept a Yacht in Port O'Conner and a bunkhouse. I liked diving with him because girls on board would fix him meals even when we were suited up on deck, and I would get one too. I made a few trips with Jim. Being a cautious diver he thought I was too hasty, he said he knew there were dangerous creatures in the ocean. Could be, I just loved spearfishing and usually the first guy there gets the best fish. Jim was a womanizer, and his first wife shot him and divorced him with Percy Foreman as her lawyer. He survived and years later his second wife did a better job putting 7 slugs in him. Jim once tried to sell me a car, I turned him down but one of his clients bought it and used it to rob a bank to raise money to pay for it. Jim defended Woody Harrelson's father Charles Harrelson who murdered a federal judge. Always on the hunt, he once asked my wife Jackie if she would like to meet him for a drink.

Jim Tatum crooked lawyer and womanizer

A TRIP TO CORPUS CHRISTI

In 1974 I was now a vice president in our growing company and finally making real money. We had business in Corpus and while I was there I called Jimmy Clark for a dive trip. Glenda Causey showed up for the trip. She was divorced with a small child. I was already in love with her. I invited her to Freeport for a trip to the Flower Garden. She came and we bonded. I told Jackie, enraged she took the boys and moved back to Corpus.

I felt I could not stay on the Texas coast with an enraged ex-wife. I told her she was still young and beautiful and needed to find a man that smoked and played bridge. Yes I was selfish.

On my last conversation with Louis before leaving for St Croix I told him he was wasting his time trying to float the Middleground by filling it's own storage tanks. He had even added 50 gallon drums tied to hand rails. Louis was obsessed with raising this boat, he was spending every penny he could get his hands on. When I was in Jr high I devoured books reading constantly. One of the books was by Navy salvage diver Commander Edward Elsberg. The Navy would prepare barges with fill and bleed valves, float them out to a wreck and sink them on the wreck. Hard hat divers would use fire hoses to tunnel under the wreck and feed cable through to cradle the hull. (I couldn't do that, claustrophobic) Once the cables were attached to the barge, they would pump the barge with enough air to float the vessel off the bottom and drag it to shallow water for non deco diving to finish the salvage and tow the wreck to dry dock.

Later while living in St Croix, I heard that's exactly what Louis did. He got hold of a RR tank car, cleaned it and fitted it with valves. Towed it out, sank and fastened it to the Middleground. Then pumped air into the tank (a little too much) it rose too quickly almost under the Aqua Safari and scared them all badley.

CHAPTER 3

St Croix

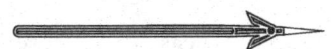

I saw a job for Electrical Supt. for Hess Oil in StCroix USVI. I interviewed, got the job and Glenda, daughter Kim and I headed for a new life. Yes, I walked away from a vice president's job and dealing with people I knew and liked to an uncertain future. The year was late 1974. The audacity of youth. I turned 40 my first year on the island.

Hess was an autocratic company. It was like you had joined the army and you were at war. Their primary management tool was firing. Electrical was in crisis, in fact the refinery stayed in a crisis management mode. Leon Hess flew down in his gulfstream most weeks. I had to sit in the morning meeting with the managers where problems were brought up and fingers pointed. It would have been comical if it wasn't my new livelihood. I found out who the empty seat next to me was for. Leon sat next to me when he was there. He was all business. I started solving problems. We lived in the camp. Later deciding to stay there because of the crime. Hess learned to love my technical skills, but hate my love for spearfishing. I started classes to train my best electricians so I could be free to enjoy my spare time. I brought a spare inverter to the classroom trying to teach them about synchronism and static switches. I finally realized I only had 2 men capable of understanding solid state inverters. Hess shift supt. would have security wake me at night or call me to come in while we were out for an evening. I had to get this under control. After a full year I was the first supt. to get up and walk out of the bldg at 5PM and not come in on Sat or Sun. The sad joke was HESS stood for Holidays, Evenings, Saturdays and Sundays. Luckily refinery people did not understand my craft and I was pretty much left alone as long as I got good results. Their solution to any technical problem was to call a consultant, I was my consultant. This was strange to them, we were just hired hands. Hess had nice secure camps with double-wides for most and a few homes for top management. Robberies and rapes were weekly occurances around the island. Tourist hotels had to maintain good security.

1975 was the year PBS broadcast a 12 part series The Ascent of Man. It was developed and narrated by Jacob Bronowski. Each segment was filmed on location in different parts of the world. It changed my life and made me love science. He later wrote a small book Science and Human Values which was required reading at MIT for several years. A book was developed from the series. I have reread it several times since 1975. You can find episodes today on you tube. The one he did in Auschwitz will break your heart.

Hess had shipped my boat down. I met a local contractor, John McCallum. We became frenemies. His wife Kathy and Glenda remain friends today. In my rare spare time I explored the island. The bay East of the Hess jetty was full of lobster. I could drive my truck down after work and pick up 3 or 4 nice bugs in less than an hour. I made a snare out of ss wire and a stick. I could release egg layden females. I learned to use the stick handle to tap the bugs into position so I could snare them. Trying to grab them by hand usually resulted in broken antennae and lost bugs. Our neighbors had gardens, so we started trading lobster for veggies. Sadly that bay is dead today, caused by runoff from Sunny Isle and maybe sewer leakage from Hess.

McCallum had a small boston whaler that I'm sure over the years drove my kidneys up under my shoulders from pounding. John had one speed. We went diving a lot with John and Kathy collecting shells, starfish and such. I hated my boat. It was always breaking down. I found a buyer and got rid of it. I bought a zodiac with a small evinrude, perfect for local exploring and going to Buck Island. (everyone's Sunday hangout) We had Kim (Glenda and Billy Causey's daughter) snorkeling at a very young age. I saw my first string bikini on Buck island, as I turned for a better view of a well tanned derriere, Glenda jerked me back around.

Scuba diving with Glenda one day, I stopped and handed her my snare and gloves. She looked puzzled until I pulled my shorts down, got comfortable and took a dump. I am very comfortable in mother ocean.

We joined the local dive club to meet local divers. We met Don and Judy Hinkel. Don became my main diving buddy, and another frenemy for life. (hills and valleys of friendship) I am a loyal friend. Jimmy Heidland and Jim Rio were loyal friends. A disloyal bad act against me makes for a strange friendship.

RED HOOK MARINA - JOHN HARMS

It was hard to know where to add this section. Johnny Harms was a boat captain for the Rockerfellers. He developed Red Hook Marina on St thomas. It was the jumping off place to fish the famous North drop, which he also made famous when he had his own charter boat, The Star Trek 2. He was on TV with Kurt Gowdy and a few movie stars catching marlin.

We chartered with Harms 2 years in a row with the McCallums John and Kathy. The first trip I landed a 50 lb YF tuna as the seas were building, a while later we found ourselves in a tropical depression. His boat is a custom made wooden boat. All we could do was idle and face the waves. This area is where hurricanes can start. As we lifted and crashed back down with huge waves I began to think about this wooden boat breaking up. Kathy and I got very sea sick. I tried to heave over the side but the deck hand Spike grabbed me by my belt and sat me down in the fighting chair with a bucket. After it layed down a little we headed in, no marlin. Kathy was feeling better when Glenda offered her a sandwich, I then heard a loud screech "it's garlic" as Kathy heaved again.

The 2nd year John and Kathy had to back out. It was a good trip with Glenda and I each catching small marlin.

TEXAS VACATION

For our first vacation we were still Texans and I had never missed the annual Freeport Fishing Fiesta July 4th. I had the record grouper 321 lbs, which is still listed every year because it is now illegal to shoot them. Both Jim Rio and I have shot larger, but not during the fiesta. In fact after we left for St Croix, Jim landed a monster 600 pounder.

We all went spearfishing and I lost at least a 400 pounder for sure because I was lazy. My slings were getting old and weak and I was too lazy to change them. I had him twice across the rig before the tip pulled out. One technique I had developed was to release my rope while riding a fish as it doubled back under or around a piling, and grab it again. Jewfish were usually hesitant to leave the rig. I got 1st place cobia in hook and line with a 72 pounder. That record was beaten a few years later.

The story of my 321 lb record fish is crazy funny. Because of prevailing winds and waves, we used to launch 40 miles down the coast at Port O'Connor. Then with following seas we would work our way back to Freeport without pounding, jumping rigs along the way. On this day in 1973 we were nearing Freeport and there were 3 jewfish in the boat, none of them mine. I was gloomy and down in the dumps. We came to a rig and I was first in, found a nice one and shot it. When it slowed down and I could turn its head and control it I grabbed it in the eyes and swam it to the boat. Jim Rio and Richard Speed saw me and I handed my rope to them and swam around to the stern to get in the boat. When I got in Jim and Richard were jumping around empty handed. "Where's my fish I screamed - you didn't lose it did you?" They nodded yes. They said it went crazy and jerked away. I went berserk and quickly put my fins and mask back on and dove head first over the side, swam straight down and saw my rope, grabbed it and drug my fish back up to the boat and handed it to Jim again. A story in the local paper stated "he had to land his fish twice"

Years later in Angleton, the local paper interviewed me about that fish again because it is still listed every year during the Fishing Fiesta. I went into detail about how we used to rig up for JF. I ended with, they are excellent food. The reporter embellished,"he enjoys JF today". This brought me a visit from the game warden. I had my original statement in writing. It was strange sitting in our front room talking to 2 armed game wardens about something that happened 40 years ago. They told me my story had been distributed all over the gulf coast and I had given out a lot of information, bad people could use to break the law. We weren't amused and complained to the local paper, they hung up on us.

321 lb jewfish

Back at Hess Oil we had child support to pay so Glenda went to work in accounting. Hess was good about hiring couples because turnover was huge, and they discouraged mixing too much with locals. The only thing you were allowed to accept from a contractor was a pen with their name on it.

CANE BAY

We were still doing some scuba diving. I had a 100 cu ft tank I had used on the Middleground salvage job. I had heard a lot about the Cane Bay drop off at 200 ft.. Thomas and JoAnn McDonald were visiting us and we went to cane bay. I decided to go down for a quick peek over the side of the notorious drop off. With no BC (rarely used the one I had) just a tank and weight belt I worked my way down to the ledge, feeling very heavy as I reached 200 ft and the precipice. When I peered over into the abyss I freaked out. I felt so heavy I feared I would fall over the side and sink forever. (narced and no BC is not good) I literally crawled back up till my senses cleared and I calmed down. No one had any idea I had just scared the shit out of myself.

Black coral shark

LANG BANK - I TRY DROWNING

Lang Bank runs out from the East end of St Croix averaging 60 to 80 ft deep and dropping off on the North and South sides to great depth. We were with McCallum out on the bank scuba diving. I had shot a grouper under a ledge and was trying to get it out when my air ran low. We were in

80 ft of water. I returned to the boat but I had to leave my gun. John had a spare tank with 800 lbs in it, I felt that would be enough. This tank had a 300 lb reserve valve that required the diver to pull it down to enable the final 300 lbs. But it was broken and the valve handle just spun around. I grabbed it anyway and in a hurry headed down. I found my gun and fish and struggled to retrieve them. Suddenly my air shut off. Didn't bother me, I had run low on air before, when I swam up I could suck a little more air. I continued struggling and finally drug the fish and gun out and started up. I sucked on my reg. Nothing, I may as well have sucked my thumb. I kept swimming up as hard as I could, now with a feeling of extreme dread. So sad, you idiot, you have drowned yourself. I let a little air out as I rose still wary of embolism. Just before breaking the surface, I needed air so bad I tried to breath water. Glenda said when I broke the surface I gasped so loud I sounded like a whale sucking air. I have no idea why I didn't pass out and drown. My throat must have locked shut when I tried to breathe water.

CHAPTER 4

Anegada

R eal free divers

When I started diving with Don Hinkel, Ronnie Cutler, Dave Coston, Dick Malpass and Slim Francis I expanded my snorkeling ability. These guys didn't mess with tanks for spearing. The more I went freediving the more I loved it. I got better and better. I became part of the St Croix team. Over the years contests called shootouts were scheduled in Culebra, Hans Lollik, St Croix, and Virgin Gorda and Anegada in the BVI. (yes we had a shootout in Virgin Gorda). Stupid. This probably helped them decide to outlaw spearing in 1988.

Everyone talked about Anegada. A 17 mile sand bar surrounded by the most beautiful reef in the caribbean. With the McCallum's and others we chartered a flight and made our first trip to Anegada. We rented a room in the village from an elederly couple, the Wheatleys. They were so sweet, breaking out fine china to feed us. We rented a vehicle to explore the island and discovered a fledgling hotel run by Lowell Wheatley and his wife Vivian. The Anegada Reef Hotel is run by their children today Laurence and Lorraine. Over the years we got to know Lowell very well. On this trip we paid a local to take us diving. I hate to say it but I shot a huge tarpon. Not food. I never did that again. I vowed to only shoot food from now on.

FIRED

McCallum was the largest contractor on the island. He had people both in Hess and Martin Marietta (a bauxite plant next to Hess) He was constantly moving people back and forth. Hess claims he was charging for people that were not in the refinery. They went through his books and found my name. McCallum had bought us tickets for a trip to Tortola. When he bought the tickets I had no idea my name would appear on his books. We bought John and Kathy meals to compensate but we didn't keep books. I was given 9 months severance pay and expenses to move back to Freeport. I rejoined the company I had helped start BES. I was now a hired hand. The guy who had taken my place 3 years ago (Tom) had become wealthy. Huge house, country club, expensive cars etc. The company was booming. Thad Brown had stepped aside and let Tom run it. Thad like to sell. He set others up to run things. (a fatal mistake) A switchgear fire at Vidriera Monterrey in Monterrey Mexico (glass manufacturing plant) needed our services. I was sent down to run the job. We had a company plane, a Cessna 360 (like the old Sky King Show) Tom was a pilot, he flew us down. I got the flying bug.

I was trained by a crop duster instructor in a J3 cub on a grass strip in Angleton Tx. After solo we went to Cessna 152 and 172. I loved flying. Like spearfishing it is addictive. While back on the Texas coast I was able to visit my boys and see how they were doing. My oldest, Stan said with the monthly check we sent they had enough money and Jackie was going to remarry.

I didn't like the situation at my old company. John McCallum had started a new company and had a big job at Martin Marietta and really needed an elect. Supt. to run it

We did a lot of spearfishing while back in Texas. Once in Corpus diving with Jimmy and Luke, Jimmy was instructing Luke on something on the bottom at 60ft. Eager to to show off my freediving skills I swam down behind them and grabbed Jimmy's leg. He whirled around in surprise. Later using a small gun of Glenda's, I shot a small dog snapper and somehow got my thumb in it's mouth and a canine through my thumb nail.

WINTER TRIP

Jim Rio had a new boat and was waiting for canvas spray shields to be installed. We got together with Cliff Woodiel, Tim Rio and Don Bickham and took off for a rig 30 miles out. It started out fun. The rig was full of winter warsaw. We had a few in the boat when I came back with my last fish. Don Bickham handed me his gun and said get one more. There were smaller brown grouper at this rig I had never seen before. I decided to shoot one. Turns out they were gag grouper, common in Florida. I liked them for food better than warsaw.

The Norther hit just as I was boarding with the last fish. We had wet suits but with the freezing wind and 4 to 6 ft sea they were not near enough. Every wave sent spray directly on us. Somehow Tim Rio was able to sit and steer directly into the sea back to Freeport. Cliff and I lay on the deck puking and freezing, while shivering violently. I believe if I had a gun I would have shot myself to stop the misery. We survived the most miserable boat ride of my life. Stupid is as stupid does. If your going to be stupid, you gotta be tough.

BACK TO ST CROIX

McCallum ended up moving us back to StCroix in 1979. We were stranded in Miami by hurricanes for a week. Arriving in St Croix Hess needed help with hurricane damage. We ran into Hank Wright (Leon Hess's man in charge of the refinery) at Frank's steak house. He told McCallum I was a "damn good electrical man" and he needed us to help get the refinery back up.

We ended up buying a house in Judith's Fancy. I tackled the Martin Marietta project. Things were not going well for my buddy Jim Rio in Texas. He and Robin came down and lived with us for a few months. I had a 25% piece of CIE (John started a new electrical company for me to run) Cates Industrial Electric.

After a year of chaos I approached Hess for a job. I was hired back as construction Supt. I had no problem with walking away from CIE. Glenda and Kim had walked in on our house being robbed and fled. We moved back into the HESS camp for security.

I started renting airplanes for trips to Puerto Rico and Anegada.

Our front yard Judith's Fancy St Croix. Reef sharks ran ne ashore

Had to climb a cliff in my socks

OCTOBER 1980

Don Hinkel and I decided to make a week-long trip to Anegada when I had vacation time. Texas had worn out and we no longer felt the need to vacation in Texas. Don had a 30 something ft Donzi. We put my rubber duck on the deck and made the 60 mile trip to Anegada. The weather was perfect. We would have breakfast and hunt fish. Come back to the Donzi for lunch and a nap, head out and hunt more fish. We rarely had to dive deeper than 40 ft. It became routine looking in cave after cave, shooting a dog snapper or a grouper returning to the boat to fillet and store fish. Saltwater baths work great. Jump in and get out. Wipe down with liquid soap, jump back in and get out and dry off.

One day I was actually getting tired of looking in empty caves when I looked in one that was open on the other end, before me was the outline of a large beautiful grouper. I got so excited I started hyperventilating and had to come up and calm down. I rested a bit, prepared my gun and dove back down, placing my gun in the cave I looked in and shot. It was a beautiful 29 lb Nassau grouper. From tired of looking to pumped in a heartbeat. It was dark when we got back to the boat and we got a poor picture of it before we fileted it. Carl Butler accepted it and I had a North American record in freediving for a few years. After our ice chest was full it seems more fish showed up. I cornered one poor Nassau that seemed terrified but didn't bolt away like the yellowfin did. I actually touched it with my spear tip. On the way home we stopped at Pajaros Point on Tortola to look around. I found another very nice Nassau in a cave. I thought, I'll get you next time I'm here.

Don Hinkel had the yellowfin grouper record of 32 lbs. Over my years in Anegada I landed 29 and 30 pounders 3 times but could never break 32.

VIRGIN GORDA SHOOTOUT - SWB

We had our annual shootout against St Thomas in Virgin Gorda. I was diving on John McCallum's little whaler with Slim Francis. John is not a free diver. Slim is a St Lucien, very dark, about 5'2" and all muscle. We were off of Pajaros Point working at 75 ft. Slim pointed at a grouper, I did a good breath up and dove. When I got to the coral head and peered around to shoot, the fish crossed the open sand to another head 40 ft away. (you never dive straight at a fish) I wasn't good enough to follow it and started up. Slim acted irked I didn't get a shot and he took off after it. When he got to the head it had moved to, it crossed back over to the original head where we had seen it in the beginning. Slim crossed over to the fish on the bottom and shot it. I'm thinking "damn Slim is good". I started swimming away and for some reason I looked back. Slim was sinking in a relaxed posture with his shooting line, gun and fish wrapped around his leg. I swam back quickly and handed my gun to McCallum and made a quick dive straight down to Slim. I reached his limp wrist about 30 or 40 feet down, thankful he wasn't any deeper and started up with a sleeping body. McCallum is a big strong guy. I handed Slim's hand to him and he lifted him up and draped him over the side on his belly. Water gushed out of his mouth. We rolled him over and Glenda was preparing for mouth to mouth when he coughed and woke up. Slim was very subdued from his usual constant chatter. He was foaming at the mouth for some time. We were later told we should have taken Slim to first aid and not continued diving.

My time is not correct, but when I was fired Slim had a beautiful mahogany fish carved for me as a going away thank you for being president of the St Croix dive club and saving his life. It hangs on the wall in front of me as I type..

The St Thomas team was composed of Carl Butler, Mark and Kevin Marin and sometimes Pat Bailey, John Graeser or Sid Smith. Carl and Mark had the loudest ego's of that tribe. Yep, those were the ego days. We all used arbalete type guns with only shooting lines. I had won 2 Dacor's in Freeport and was still using them.

N43723 - THE ANEGADA WARRIOR

Glenda found a good deal on a 4 passenger Piper Warrior. We became airplane owners. It was friendly and easy to fly. We bought a good inflatable for emergency water landings. Thus began 5 years of adventure with our own bird. The drill for Anegada trips was clearing customs at Beef Island, then buzzing Lowell and Vivian Wheatley at the Anegada Reef's Hotel, then buzzing the runway to clear the goats off and finally landing. Lowell would send a truck to haul us and our gear to the hotel. I had gotten my Zodiac over several months earlier by renting my friend Bob Dunn's Cessna Cardinal. He used it to haul materiel to his job sites in Puerto Rico.

I had adopted Kim by now. She liked to snorkel but never really got into spearfishing, choosing horses and Pony Club instead. Kim and Glenda became very active in St Croix Pony Club.

The Anegada Warrior

DON HINKEL - FRENEMY

So many trips, they start running together. Don and I were working our way down the Anegada reef one day when I saw him look under a ledge and shoot. He came up empty handed and very excited. "There are 2 groupers under that ledge he said, give me time to reload and I'll give you first shot" I believe my friends. Waiting until he was ready, we both dove. Arriving about the same time, Don quickly shot. I grabbed the edge of the ledge and peered in and shot the 2nd grouper. Two nice yellowfin. I never felt the same about Don after that. My word is my bond and spearfishing is serious fun.

The air force had a base in Puerto Rico, sometimes while snorkeling the reef their jets would fly 100 ft over our heads, quite a dramatic break in the silence.

Yellow fin grouper

cubera

Yellow fin grouper

Anegada Reef Hotel pier

Nassau and Yellow fin grouper

GLENDA AND THE SHARK

Glenda and I made a few trips to Anegada by ourselves. On one trip the ocean was so calm it resembled a swimming pool. We spent the whole day circling the entire island. We never saw another soul. We stopped and napped at noon at Loblolly bay. It was so clear you could see fish from the boat. I decided to only shoot one fish. I spotted a nice yellowfin. I made my dive away from it and on the bottom worked my way back and peered over the coral. There 20 feet past the YF was a Nassau way bigger. (Nassau are much better food) Back to the surface, long breath up and dive again for the Nassau. We got in after dark, they were about to come looking for us.

Next day we just stayed in a small area where there were some dog snapper. There was a rather large reef shark doing his thing cruising the reef. You don't mess with reef sharks, they are territorial and if you try to run them off they will attack. I watched it swim down the reef and asked Glenda to get in and warn me if it came back by hooting in the water. I spotted a nice dog snapper. I made my dive but the snapper left its cave and crossed the sand to another part of the reef. I felt good so I followed it, peeked in and shot it. When I got to the boat Glenda was in the boat and pissed. To this day she will tell you I swam off and left her with this big shark.

Another friend I dove with there was Bill Batts. Bill finally got a grouper, and was so excited he was telling me all about it as I kept pointing down and poking the shark trying to get his fish. When he finally looked down he tried to swim backwards over the reef.

I have observed these sharks swim over the top of the flat part of these reefs, being caught and stranded by the ebb of a diminished wave. Stranded they simply wait for the next wave to lift them up and continue.

ST THOMAS SHOOTOUT

A local dive shop owner Bret Gilliam was a busy entrepreneur. He was active in Skin Diver Magazine. He had a beautiful charter yacht built in Main. We set up a shootout with St Thomas at St Thomas and Bret ferried us over. We would split up and dive with 3 of our competitors. The rule was spanish mackerel had to be over 1 lb. I got into lots of small macks and brought in 4 or 5 hoping some were over a pound. None were. Bret, always having fun, tried to talk them into giving me points for accuracy. Since I was now into free diving, I lost track of Bret and his technical diving, charter boats and magazine career in which he became wealthy.

I'm not sure which trip but I was out late in the day with a St Thomian diver Sid Smith. We were just training and freediving around Hans Lollik.

The bottom had a gradual slope down to 75 ft and then a steeper slope down much deeper. We were just breathing up and doing 75 ft. I became relaxed and comfortable, and after a long breathup I descended to 75 and was so comfortable I spotted a coral head further down the steeper slope and swam down to it. I know I was over 100 ft, I looked up and scared myself. The surface looked a mile away. I don't think I ever shot a fish below 75 ft because I tried to stay in my comfort zone. Years later Terry Maas preached the same thing, stay in your comfort zone, too many have been lost exceeding the ability of their body, including Terry's son.

THE PEQUOD III

I had talked John McCallum into buying a Classic Seacraft 23. I had ridden in one during a St Thomas shootout, I was very impressed. It was "the small boat" of the 70's, winning the Miami to Bahama race in its class a few times. John eventually wanted a bigger boat and bought a large Boston Whaler. I bought my 3rd boat from John. Both Mark Marin and John Graeser owned Seacraft. I was now maintaining an airplane and a boat besides a high stress job. Youth and energy are great.

STANLEY'S REEF

McCallum was dragging me behind his whaler one day when we ran over a beautiful reef in 60 ft of water on the South East end of St Croix. There were fish there and a group of reef sharks always on patrol. I came back often in the zodiac off the beach with Slim Francis and once with Ronny Cutler. I warned Ron about the sharks. On his one visit when I returned to the boat Ron was there very upset. He had shot a fish and had to abandon it to the sharks.

Slim came one day with a float line. His plan was to shoot a fish in a cave, surface and pull it up. It went as planned till he started pulling it up. I saw my first feeding frenzy. After the fish was devoured a late comer showed up and bit the now bare shaft. I never felt threatened laying on the surface watching. Slim advertised the reef to friends and St Thomas, it became known as Stanley's Reef among our friends. When we returned to the reef after the 79 hurricanes most of the beautiful coral had been pounded down, very sad. It would take a thousand boat anchors to do the damage of a hurricane. I am not, of course, in favor of dropping boat anchors on live coral.

LOUIS AND GLORIA

I received word of a tragic event. Louis Shaeffer and his wife Gloria had a tumultuous marriage. I was caught in the middle for a while being close to both. Glenda and I had lived with them for a short time before we went to St Croix. They finally divorced with Gloria staying in the house at Bridge Harbor Marina. Louis showed up one night and wanted to see Gloria. Her son wouldn't let him in. He insisted and became aggressive. The boy went and got a shotgun and told Louis to leave. This led to the boy shooting through the glass door and severely damaging Louis's leg.

Louis ended up losing his leg just below the knee. Only a few years later, Gloria died in a car accident.

1986 - MOVE TO ST THOMAS

Hess had one of the largest tank fields in the world when the price of oil crashed in 86. I had just completed a large project of computerized oil proving and tank gauging. I was laid off with 9 months severance pay. The job market in the states looked bleak and McCallum needed someone to put in a co-gen plant at the new Virgin Grande hotel in St Thomas. We rented a condo in Mahogany Run with a Northerly view of Hans Lollik. We enrolled Kim in Antillies the local private school, Mark Marin was headmaster and his brother Kevin a teacher. When Hess hired me back 9 months later, we pulled Kim out and enrolled her back in Country Day in St Croix. This upset Mark greatly. He depended on these tuitions to pay his salary. (which he was quite proud of) I once told Mark I was the best electrical man in the USVI, he then asked me "yeah, but are you making the money of the best in the VI?"

John paid me cash (under the table - I ended up making very good money that year. I found a place for the Pequod 3 in a marina with forklift boat launching. Of course moving is always a pain. There is much better diving in StThomas than St Croix. We took full advantage of diving with the Marine brothers,Carl Butler, Sid Smith and John Graeser. We had easy access to St John, Hans Lollik, and Culebra. Another cool spot there is Sail Rock where Sid Smith shot a record 7lb yellowtail snapper (one of the hardest fish to approach and get a shot at) Our airplane came in handy running the 40 miles back and forth to St Croix.

Nine months later I wrote a letter to Hank Wright CEO of Hess Oil, explaining all the projects I had completed (including corroding fire water lines, docks and tank cathodic protection) Cathodic protection is actually boring to an EE but corrosion mitigation is very important and I threw myself into it night and day until I was satisfied. My old boss the construction manager was made maintenence manager, then he and Hess hired me back with a 20% raise.

BACK TO HESS - 1987

We picked up where we had left off and tried to find a normal life (whatever that is) . My oldest son Stan came down and I got him a job. After a few months he got homesick and went back to Texas.

Glenda and I decided to spend a week in Anegada, crossing in our Seacraft 23, the Pequod 3. These crossings were not easy with the average seas running 3 to 5 ft from the East. We caught a good forecast of 2 to 3 and took off. We took groceries and rented a little house. The wind and seas picked up and diving was a problem. This was not a fun trip. We tried to pick a good day to run home and finally left with scattered thunderstorms around. In Sir Francis Drake Passage we pulled into the harbor of Peter Island to wait out a rain storm. There was a pretty reef on one end of the harbor so we went snorkeling. Later sitting in the rain with our wetsuit jackets on, I watched a sailboat motoring into the harbor, coming over from the Moorings on Tortola (a popular sailboat rental spot for tourists) I watched in horror as it approached the reef we had just left. Reef is hard to see in the rain. Many yachtsmen wait for an overhead sun when in reef areas. I stood up and screamed "starboard" as loud as I could. They veered right just in time. As they passed in front of us, they wouldn't even look our way or thank us for saving their vacation. We had a very rough 40 mile ride to St Croix. Glenda is a saint, married to a crazy man.

BARBUDA

John McCallum had a twin prop company Skymaster and a pilot to fly it. We organized a trip to Barbuda. You have to clear customs in Antigua and then hop over to Barbuda. Hiring a local boatman was no problem.

Somehow we hit a day when grouper were everywhere. We had a blast. The locals were all walking around holding radio's by their ear. We found out they were listening to a cricket match.

A few weeks later Don Hinkel and I planned another trip to Barbuda in his new cessna 182. We took groceries and ice chests and planned to stay a few days. The Antigua terminal was busy and we ran into 2 attractive Canadian tourists. They wanted to come with us to Barbuda. We talked ourselves out of it. Better be good than sorry.

The spearing was not as good as last time. We moved around a lot to get fish. At one point I heard Don screaming for my gun and swam to him. He grabbed my gun and swam down and shot a dead cubera laying on the bottom. He reminded me of when I had shot my record Nassua, so excited and pumped, it was pretty funny. The fish weighed 65 lbs. The biggest cubera I got in Anegada was 33 lbs.

One side of Barbuda is tilted up. It looks strange as it rises above the flat island. Earth shifting and tectonic plates must have been involved because volcanic activity leaves a different footprint.

Barbuda

A few weeks later we made another trip to Anegada in McCallum's big twin engine Whaler. Bill Batts was along. I had met Bill while working for John and we had done some previous fishing and diving both on St Croix and Anegada.

At day's end McCallum had decided to spend the night anchored out on the reef. The 3 of us were sitting on the stern cleaning fish. Suddenly we realized the bilge pump wasn't working and we were sinking. The thought of spending the night out on the reef horrified me. I always had my canvas tool bag with me for essential repairs. I grabbed it and placed jumpers directly from the battery to the pump and it started pumping. Not satisfied, McCallum started the engines and took off in the dark at top speed. Anegada is all reef and coral heads. We headed away from the

reef into an area of sparse coral heads. I knew the reef better than anyone, we parallelled the shore and ran wide open until I saw the lights of Anegada Reef Hotel. I told John when to turn straight into the hotel, praying we wouldn't hit a coral head. I still can't believe we made it safely.

BVI OUTLAWS SPEARFISHING - 1988

With spearfishing outlawed I would have no use for my airplane. I sold my bird to the comptroller of Pusser Rum in Tortola. I agreed to deliver the bird. It was a very hazy day as I took off for Tortola and Beef Island. The haze was blocking my view of the ocean. After a while I began to hyperventilate and panic. I was going to die. I pushed the yoke forward and dove towards the water. As soon as I saw the water I leveled off, calmed down and was fine.

I bought a book and studied this phenominum. When our brain is deprived of our visual well being, this brings on stress, the muscles in our neck tense up and begin to limit blood flow. Our brain says this fool is depriving me of oxygen, I may have to faint him and relax him so I can get more oxygen.

This is the same reason relaxing is so important in free diving. It takes many hours of instrument flying to be able to relax and fly blind. I don't think I could ever do it. Yes, you can believe your instruments, but your brain wants to know your well-being at all times. It's not a computer game when you can die and kill those with you, the stress is very hard to overcome. Weather is the greatest killer of good pilots. (and sadly their families)

LAST ANEGADA TRIP

Even though spearing had been banned, Don Hinkle and I decided to do a trip in his latest cigarette boat to my beloved Anegada. Louis Shaeffer had called me earlier wanting to borrow my boat and take a new client to Anegada. There is no way I would ever loan my prize possession to anybody, much less Louis. Louis always needed something, he was always needing help. Back in the Middleground days we used to try to avoid him if we weren't ready to help him, he would start with, we this and we that like we were on his payroll. I think he had started at least 4 diving companies that had gone bankrupt since the Middleground days. (but stay tuned, he ends up wealthy)

Don and I were working our way down the reef looking for grouper and snapper. I was using my Dacor aluminum tube gun (no reel) when a monster mackerel swam by me. I made a shallow dive and couldn't believe it turned right in front of me. Thinking I might lose my gun, I swam towards it pointing and shot. You achieve your most accurate shot when you point,(not aim, point) and swim towards the fish and shoot. I couldn't believe I stoned it. As I swam it to the boat holding one gill in each hand, I prayed it wouldn't wake up. Don came back to the boat and freaked out at the size. Back at the hotel we didn't bring it ashore, fearing law enforcement. Louis came out to see it. Having no scales we guessed it in the 50 lb range. You can see it today on my Facebook page.. It will be there when I die. It is the fish I am most proud of. (well, there have been some big red snapper)

Large king mackerel

SEPT. 17 1989 HURRICANE HUGO

The summer of 89 I had a run in with my boss. He jumped me about my men quitting early while I was designing a fix for a problem (75 electricians in my dept.) I was now a general supt. With 2 supts. And 6 foreman. I was tired of him running around scared and looking for anything wrong, he was Hank Wright's on site hatchet man. I lost my temper. I wish they would have fired me (we would have missed Hugo) but they put me in charge of refinery electrical expansion because I was such a good engineer and electrical problem solver. I was always doing small projects to improve the electrical system besides maintenence, instead of bird dogging my men. I now had a wonderful job, same pay and no people under me to worry about.

We got a phone call from Jim and Robin Rio informing us we were in the path of a monster hurricane. A week earlier I had filled our freezer with wahoo trolling on Lang Bank. The wahoo appear in September every year.

We rode the storm out in a brick control building. People sought shelter as best they could. There were many stories after the storm. We had sustained winds of over 200 mph. There was massive damage. The refinery actually came out pretty good. We gave the wahoo away to anyone who could cook it.

The local military and police joined their relatives in looting the island. The 101st airborne was parachuted in to restore order. Later the local national guard general was fired.

Eastern airlines had just gone bankrupt. Hess pilots knew some Eastern pilots so Hess chartered an Eastern plane to fly in and evacuate our woman and kids. Glenda described a black lady soldier that escorted them to the airport. They were stopped by local staff who said they weren't getting on that airplane, they will stay and suffer with us. The soldier thrust her weapon into the man's chest and stated, "these people are getting on that plane" There was no food on the plane but there were plenty of small bottles of booze. There were some tipsy people on board when the plane landed in Miami. Thomas and JoAnn McDonald met them at the airport in Houston.

I spent a month there after the storm, helping restore power and salvaging as much as I could from our new doublewide. We actually came out pretty good. I gave our furniture and hot tub to McCallum. I still have a stark memory of his trucks driving away with our "stuff". It was declared a total loss and we collected insurance on our furniture. I didn't have time to mess with the new elect. Supt. and went about restoring power on my own. This upset him greatly. I was at the admin bldg. repairing a transformer when Leon Hess pulled up with Hank Wright. He sat in the car and cried. This refinery was his pride and joy. He had it kept very clean with air pollution very low. (we won't discuss leaking sewer pollution) Visitors would ask, "Why is the refinery shut down" when they drove by, but it was running full blast. They expected to see pollution from our smoke stacks.

A few years before they had given John Hess some money to hold management training at the refinery. We had to watch 12 O'clock High as part of our training to manage stress. I explained to John the wrong people were in this class, he needed to get his dad and Hank in here, they rule from the top down in a constant crisis mode.

Glenda's arthritis had gotten so bad I had just gotten her a wheelchair before the storm. A Puerto Rican doctor had her on radical meds including gold shots. She needed a real rheumatologist. I resigned from my job and prepared to start life over for the 4th time at age 50.

CHAPTER 5
Dow Chemical Designer/Engineer

We moved back to Alvin Tx and stayed with Thomas and Joann McDonald until we found an apartment. My old company had been sold to a huge electrical entity. Thad Brown had been run off by Tom and others with stock who joined Tom against Thad. Tom didn't want to pay taxes and became an anti govt. guy taking his few million out of the country somehow, under threat of law enforcement. He left his wife and children and started a new life in El Salvador.

They remembered me and I worked there for a few months until I got on with Dow as a designer.

The Anegada Warrior On the way to Puerto Rico here

The Pequod 3 had beat me home. Construction barges from Houston were at Hess empty after delivering construction housing. I got permission to pull my boat onto a barge for the trip back to Houston.

We bought a home in Alvin. We were 50 miles from Freeport and offshore fishing and diving. Kim finished her senior year in Alvin. One of Nolan Ryan's sons was in her class. She said the teachers spoiled him rotten.

Living in Alvin was unsatisfactory. We wanted to build a home in the Bridge Harbor Marina near Freeport. I went to the bank and they handed me 5 lbs. of paper to fill out. John and Kathy McCallum had divorced. She had money and offered to finance the build. After completion we could seek financing of the finished product and pay her back. That's exactly what we did.

We now lived next to the intercoastal canal and could be offshore in 30 minutes.

Hurricane Hugo had made John McCallum rich. Being in the right place at the right time paid off big for John. He housed the government officials who came in to restore the island. They started handing him many contracts, and John knew how to make large profits. He urged me to stay and be his electrical contractor. Working for him before had been a nightmare. I told him no thanks, it wouldn't be any fun. Being selfish, I try to have fun every day and I rejoiced in my hard earned skills when I solved a difficult problem. When I was putting in the oil movement system at Hess I used to wake up at 3 in the morning and go in the refinery. 4 hours of sleep seemed to satisfy me. Still, Hess was never able to paint me yellow and green. I would have my life. I thought about writing a book about Harsh Oil and Leon Harsh and Hank Wrong and their sycophants. The locals pronounced Hank, Honk, which was pretty funny to hear.

HALF MOON CAY - NEAR GUANAJA IN THE HONDURAN BAY ISLANDS

Jim Rio's brother Tim and wife Lisa had gone to work for a retired plastic surgeon in the Honduras Bay Islands. Jack Conlee had lost a lawsuit for a million bucks on a titty job. He turned around and sued the manufacturer and won 10 mil. He then retired and bought a small island, Half Moon Cay. Jack and wife Darian hired a crew of locals to build a home out of Honduran mahogany for them and their companions Terry and Diane Smith. Tim was hired to supervise the crews. Two guest houses were also built.

Tim and Lisa were allowed to have guests, so his 2 brothers Jim and Mike Rio made trips down for vacations. We found out about the grouper spawn and what time of year it happened. They invited us down to see it. We ended up going 2 consecutive years in a row to try and see the spawn, and of course spearfish. This was the early to mid 90's and security had tightened up. I labeled my speargun "fishing gear". It was an interesting trip. You fly to Tegucigalpa then catch a DC3 over to Guanaja where Tim and Lisa picked us up in a boat. They stacked chicken cages and suit cases in the back behind the passengers in the old DC3 (nothing to keep it all from flying forward in a crash). Our first trip was scary. We were in a rain storm and the pilot was flying 50 ft. above the water so he could see, we were hoping he would find a runway.

It was a beautiful island and our hosts were very gracious. Tim was not a spearo, he and Lisa were sailors (they loved sail boats) I grabbed my gun and with Tim took off to see the reef. Tim had never snorkeled completely around the little island, so that's what we did. I didn't shoot anything but I knew fish were there.

Next day I told them I would shoot supper. They laughed and said the fish were too wary to approach. Not true, I got a nice grouper. Tim wanted me to let the hired hands clean my fish, no way, I have always cleaned my fish, it's part of the trip. Lisa cooked coconut grouper and I still remember the delicious taste to this day. I spent a few hours trying to use a pole spear, but the fish seemed to always explode away just before I would release the shaft.

The spawn was very interesting. Huge Nassau grouper turn black and white as they prepare to spawn. The water was in a state of excitement, with huge snapper and a few sharks cruising around. Meanwhile, commercial fishermen were sitting in their boats catching as many as they could. From my days of reading Cousteau, I carried a stoute stick to poke any curious sharks away, a habit the Costeau team had adopted. We used scuba, of course in this situation. A biologist was also there documenting the spawn.

The Conlee's and Smith's had a picnic for us and also had us over at night to their home for drinking, dancing and hot tub. We had 2 lovely vacations at Half Moon Cay. Looking online today, I see it is for sale. Before you buy, think "hurricanes" they have bad ones.

Tecucigalpa was an interesting place. Military men with rifles on the corners. There was a mini fridge in our room with drinks in it. We had no clue. When we checked out they wanted something like $75 for their little bottles of booze. What were we to do? On the airplane back to Texas it's grab your own seat. We found a seat for Glenda, but not me. I went up and down the aisle seeking an empty seat. A beautiful blond was breastfeeding her baby with baby stuff in the aisle seat. I asked her if she had bought a seat for the baby, she said no. I told her the situation and she slowly removed the baby bags etc. from the seat and allowed me to sit down. I looked straight ahead most of the time, but I had to take an occasional glance at my seatmate. She warmed up a little and we visited. She was stunning, and I'm sure used to being admired.

THE FREEDIVE LIST

For my first design job at Dow, I had a cadder. After that I must learn to design and draw on the Dow computer system. I loved it. Earlier, we bought a laptop for the house and thus began our modern era of communication.

Jim and Robin Rio had their own boat. Jim loved to take people fishing and scuba diving. He and brother Tim even had a charter business for a few years. Jim and his dive buddies were all still doing scuba while running longer trips, shooting mostly snapper and cobia. Warsaw had grown scarce and jewfish were now illegal. When I left Texas no one was eating amber-jack, now everyone was eating AJ's.

I discovered the Freedive List started by Mark Barville in California, it was global. It was great fun and I learned many new names, many of them well known. I ran across one guy nearby looking for dive buddies to dive with him out of Port O'Connor. Guy Nesbitt. I invited him to come and dive with Glenda and I in Freeport. He showed up and he, Glenda and I took off in our Seacraft. We didn't get that much until on the way in. I stopped at a rig and was the first in. It was full of nice size red snapper. I bounced my spear off of one, and finally got another. Meanwhile Guy got in and started loading the boat with snapper, most in the 12 to 17 lb range. Guy had a Riffe gun and reel.

My rubbers were weak again and I was embarrassed, with only one fish still using my old Dacor gun. I needed a real speargun with a reel. I ordered a Riffe #2 standard and a reel. I put kevlar line on the reel and small SS cable shooting line. I now had a real oil rig speargun. I had to quit being so cheap and lazy and get serious about freediving oil rigs. I couldn't believe the price of really good spearguns. Looking back today at 86, I still have that gun and I have shot so many fish with it they all run into a blur except for a special few.

One year and for a few after, Florida pompano moved into our near-shore rigs. I had a blast shooting pompano, some as big as 8 lbs. Glenda is an excellent cook and her pompano were to die for.

My old evinrude overheated and finally failed (my fault, lack of yearly spring maintenence on the thermostats) Kim was in college so we were boatless for a few years. I did some diving with Nesbitt in Port O'connor and my old friend Jim Rio. Guy could do 85 ft. Up in years now in my 60's I was comfortable at 60 ft. and on a good day 70. Guy was an aggressive hunter.

For my 60th birthday I bought a Klein Quantum Race road bike. I decided cycling would be a great way to stay fit for free diving.

We made some new friends at Dow, Bob and Margaret Pratt and their daughter Taylor. They were avid fishermen. Bob knew all about my spearfishing and records. It was kinda funny, he said "your Stan Cates" Over time I was able to get them all in the water and get Bob to try free diving and spearing. They were also cyclists. I joined the local bike club for training rides that usually consisted of 45 miles. I ended up doing 6 MS-150's (Houston to Austin, overnight in La Grange)

The Pratt's were impressed with my 21 lb red snapper that won the spearfishing division in the Fishing Fiesta that year. I was with them when I shot it. They were fishing and I was stalking a big red snapper. I would dive and it would swim deeper and away from the rig. This went on for several dives. Finally I did a long breath up and dove away from the rig, near the bottom I turned towards the rig and moved in. The snapper started its run away from the rig but I was deep enough to cut it off and get a shot. When I brought it to the boat I let out a scream of joy. Bob and Margaret looked at each other and grinned. I loved to beat scuba divers. At the scales other fishermen couldn't believe I had shot this fish "snorkeling" I thought to myself, if I had been with Nesbitt, he would have shot MY fish on his first dive. He was an athlete, aggressive and years younger than me.

Freediving oil rigs in low visibility is basically diving alone. I have shot many many fish diving alone or separated from other spearo's. It is a rare situation where a dive buddy could save you in low visibility. Just stay calm and stay in your comfort zone. Experienced rig divers are used to poor vis shooting. As I aged and became less athletic I spent a lot of time stalking certain fish. If a young skilled spearo had been close to me, he would have shot "MY" fish. I love diving and stalking alone. Your buddy dives one side of the rig, you dive the other. If a diver is depending on someone to save his life, he is probably diving too deep, staying too long. However, I do understand diving in pairs, one up one down in open water well clear of rigs.

Once in St Thomas diving with the Marine brothers, I witnessed Kevin,clawing desperately with both hands for air, his gun thrown over his shoulder hanging in place by the shooting line. I couldn't believe he had pushed himself into panicked desperation while out having fun. I couldn't count the times a fish showed up at the end of my dive and I ascended empty handed.

VENICE LA.

Some guys on the FD List were talking up shooting wahoo at the lump in Venice La. They called it blue water hunting I guess from Terry Maas's famous book Blue Water Hunting. We agreed to meet up and one of the locals chartered a boat. Wahoo appear there every year in the winter. It is the closest area to shore that they migrate past, about 20 miles out. Years later I learned you have to run out about 150 miles in Texas to reach the migrating hoo's.

On this trip I met Marcel Garsaud, Bill Delabar, Roberto Reyes, Michael Freeman and a Doctor whose name I forget. The doc was a cool guy. He had been a military doc and had a rebreather with him. He also had 3 very cool spearguns. This was before I bought my Riffe gun. Seeing my pathetic gun he insisted I use one of his guns. We spent a couple of hours hunting with no results. I was getting hypothermic so I gave up and stayed in the boat, and desperately tried to get warm.. Finally Marcel got a nice Hoo. He was very excited and happy.

We ended the trip out on the Midnight Lump. With only one hoo in the boat, the doc put on his rebreather and went down about 60 ft. to hang out and wait. After a while we saw his float take off. We picked him up and followed the float, he jumped in and retrieved his fish. These were great guys and shared their delicious hoo steaks, we had sashimi on the dock.

At our motel I was a zombie and it took a long hot bath to revive me.

Side story. Thank god Glenda stayed at the marina, she would have frozen as I did. While hanging out she saw a very large boat being hauled out at the ramp. They were burning the rubber off of their truck tires. She walked over to them and suggested they might start their outboards and get a push. They talked it over and did just that, killing the engines just as they cleared the water. This was standard practice for us.

COBIA FEVER

A guy on the Freedive list convinced me to buy a Rob Allen gun. (my favorite today based on price and accuracy) Bill Crawford and I exchanged comments on guns and methods over several years.

Bob, Glenda and I were on a trip going from rig to rig one day. I had bought a one band closed muzzle RA gun for shooting smaller fish, like my new favorite, (mangrove) gray snapper. I made a deep dive in the middle of a rig and was slowly ascending when I felt a presence near my shoulder. I glanced over at the biggest cobia I have ever seen. I have shot and caught them in the 70 lb range, this was their daddy. I looked at my puny gun, I was inside all the pilings, no use shooting, I would surely lose it. I headed for the boat to get my Riffe hoping for another chance at it.

I was loading my slings when I heard Glenda scream. I knew Bob had shot it. I swam back to the rig and found Bob struggling with the beast, it was near the bottom. Somehow it worked its way outside the rig. I started breathing up to make a dive and put another spear in it. Just as I got ready to dive I looked down and it was coming up towards me with a slow hypnotic undulating motion. I froze transfixed. At 10 ft it turned and started back down, Bob's spear stuck in its head. I remained hypnotized. When it reached the bottom it shook the shaft and swam away.

To this day, I will never know why I didn't shoot that fish. I have read about buck fever freezing hunters, maybe the beautiful undulations as it rose towards me had given me cobia fever.

A STEVE IRWIN INCIDENT

On another day we were rig hopping in poor visibility. Glenda and Bob wouldn't even get in the water at one rig. I got in and dove down about 30 ft knowing I could look up at silhouettes and get a shot. As I looked up the shape of a cobia appeared out of the gloom and I shot. The fight was on, it managed to drag me against the rig and then I lost it. Back in the boat I realized my brand new Riffe diving knife was gone, brushed off of my leg against the rig while fighting the fish. I keep a small tank on board to retrieve deep tangled fish when freedive retrieval would be difficult. I decided to use scuba, go to the bottom and retrieve my knife. As I reached the bottom in zero vis under the spot I had lost the knife, I started feeling the bottom for my knife. Suddenly the bottom moved out from under me. Holy shit, I was right on top of a monster stingray. To hell with the knife. Back on the boat all I could think about was Steve Irwin who had just lost his life from a stingray. Tiger sharks feed on stingray and stingray barbs are found embedded in their lower jaw, they thrash their barbed tails upward to strike an adversary. One that big would have a nasty barb.

HALF A SNAPPER

One day out with Glenda we had a murk layer on top like they get in Venice and Port Fourchon La. I dropped down about 30 ft to clear water and shot a large gray snapper, it furiously headed down as I headed up. As I pulled the fish up I felt a hard jerk. When I got the fish, the tail half was gone. I hadn't seen any cuda's and it was pretty jagged. As I swam to the boat I was wondering if the shark would come after the other half in my hands. We still got 2 nice filets.

My experience with sharks was that if you could see them, they were much more cautious. Below a murk layer and out of sight, they were more aggressive.

When diving with Jim Rio and the scuba divers they rarely saw sharks because they stayed inside the rig. I had learned to mostly freedive outside the rigs in case I shot a cobia, they were bad about swimming around pilings. This made retrieval more difficult. Outside the rigs I usually saw sharks in the background on my deeper dives. However, in every case outside the rigs with fair visibility, I was always able to retrieve my fish, the sharks were cautious because I could see them.

I kept a CO_2 float attached to my gun. Once after shooting a rather large snapper near the bottom it was fighting hard and 2 sharks moved closer. I headed up and was having trouble pulling the snapper off the bottom so I pulled my CO2 float to assist me. They didn't get a bite. Sharks are always looking for an easy safe bite.

Half a mangrove snapper

THE BLUE FISH

I did a trip with Jim and Robin Rio. The current was ripping that day and I couldn't find any snapper or cobia. I was snorkeling at one rig, staying behind a piling to get out of the current. With strong currents it's best to get behind a piling and rest your legs while you breath up for a dive. Jim and Robin were fishing. I saw lots of blue fish kinda stationary facing into the current, having never shot one, I decided to shoot one. I made my drop and placed my shot and started up, pulling the fish towards me with the shooting line like I did with smaller snapper. When I broke the surface my snorkel strap broke and I didn't have a snorkel anymore. I was holding my head back to breath without a snorkel when I felt the fish, as I grabbed it , it grabbed my left bicep. Holy shit. I swam to the boat and hollered for Jim. Every time I tried to pull it off of my arm, it would clamp down and

I would let out a scream. Jim got a screwdriver and tried to pry it off but that was killing me. I finally told Jim to get a filet knife and cut its head off. As he held the knife above the head I pushed the fish towards my bicep, which relaxed the bite. By now this had been going on for a long time and I thought "what if we cut its head off and it still won't let go?"

Jim called Glenda and told her ``I had been bitten by a blue fish "where did it bite him she asked" Wimmen, always thinking of themselves.

It took 12 stitches to close the wound. Later we did a 15 minute TV reenactment on "It Happened to Me" for my 15 minutes of fame. The other 15 minutes of that 30 minute program was a guy terribly mauled by a bear, which made my "survival" story sort of comical.

The bluefish incident

DEEP BLUE WATER SHARKS

All the years I have known Jim Rio I am not sure how many boats he has owned. His first was a Formula with 2 mercruiser inboard outboard he had bought from the criminal lawyer Jim Tatum, a terrible mistake. This boat almost drove him crazy with problems. He ended up driving it into the Freeport Jetties (on purpose) while I was in St Croix. His latest was a Main Coaster, designed similar to the lobster boats in Main. High bow, nice large wide flat deck. It had 4 lovely padded deck chairs. Jim took them off and placed 2 large ice chests in their place. I made an overnight trip with Jim to the Cerveza and Tequila rigs about 75 miles out. Sitting on ice chests was killing my back. He had bean bags for running, but once you got there it was sit on ice chests or try to get out of a bean bag to fight a fish. We caught several nice blackfin tuna at daybreak.

During the day they went scuba diving for lobster and shells on the rig. I went snorkeling. The water was pretty empty, but soon a large school of rainbow runners came by. They were larger than I was used to seeing inshore so I dropped down and shot a nice one. I was immediately swarmed by small sharks. I started heading for the boat, they swarmed around my fish and my feet. I wasn't used to this many sharks. The guys were already back in the boat as I reached the dive ladder. Every time the sharks got too close to my fish, I would jerk it away. The guys were telling me to hand them my gun, but I was busy poking sharks. Finally I handed them my gun and as I climbed aboard one of my finns caught in the ladder and I thought a shark had me by my finn. Pretty funny now.

Since no shark took a bite, I guess I was safe but I really thought I might be in danger.

Back ashore I told Jim I would not go out with him again until he put some chairs back on his boat.

Not sure when but Jim was taking a coworker out in rough seas, the man fell hard and broke his back. He sued Jim. (I thought of Guy Allison, the Corpus Christi PI lawyer who never let us pay)

I told Jim if he would have had a chair on board, the man would not have fallen. He said "you sob, you will never get on my boat again" Jim is high strung. He stayed mad at me for a few years, but finally got over it.

JON DUBOIS - 63 FT CUSTOM MERRITT

We had a wealthy neighbor in Bridge Harbor Marina. He had a 63 ft. custom Merritt. I kept after him to learn to spearfish. After some months he finally took a diving course from Mike Cryer in Lake Jackson. To say John was timid would be an understatement.

I can't remember the year but one summer we had perfect weather and made 12 consecutive trips to "the intersection" about 65 miles out where ships set their course to enter Freeport harbor. There were 3 or 4 rigs in the area. The water was usually very clear. These rigs swarmed with barracuda.

John brought Mike's 2 sons along at first as safety divers to watch over him. He would always flash the ok sign at me over and over, it was pretty funny. He was scared to death. He had to wear a full face mask and refused to learn to snorkel.

The rigs were loaded with gray snapper. There were so many cuda I had to grab my snapper as soon as I could and carry it back to the boat. I never lost one to the cuda's, but I'm sure an errant snapper would have been devoured immediately. When trolling around close to these rigs you couldn't get a fish in the boat for the cuda's.

Jon was very generous and we were able to invite our daughter Kim, Bob, Margaret and Taylor Pratt and on one trip our family doctor of many years Jimmy and wife Chris Smith on a Flower Garden trip. John had bought a Riffe No Ka Oi from Mike Cryer. He was never going to use it so he gave it to me

Mike Cryer's wife Michelle became a good freediver and spearo. However, Mike made his money teaching and selling scuba.We joked about the elevator scuba divers that got in the water with air in their BC's and let the air out to descend. I thought back to my VA Fogg days when

I would fall in head first and hit 100ft in about 2 minutes. They made a couple of trips with us on our Seacraft. We got into red snapper on one trip and had a blast boating several nice ones. Mike organized a trip to Grand Cayman. Michelle was going to take a freediving class there. I heard the Riffe girls were going to be there for the class. Taking a freediving class never entered my mind. We signed up for the trip. It was enjoyable and I got to meet the Riffe girls Jill and Julie. Glenda and I did a lot of snorkeling on the no spearing reefs, it was great fun getting eye to eye with tame grouper and snapper. I got a kick out of doing a long breath up and swimming alongside the tourist submarine as well as cleaning the nipples of the angel on the bottom at 60 ft. Of course we did the stingray feeding trip, all using tanks except me. This was my first experience with the term Master Diver and I heard it used over and over.

The Riffe girls

IG COBIA

Bob Pratt bought an old steel boat with a large diesel engine. The Kate C. This became a great fishing and diving boat with a platform on the back. We made many trips in this boat for several years. On one trip I had shot a big cobia. I always let them swim down about 20 ft so they can calm down (thus the importance of a reel) while I swim to the boat. I thought it would be funny to hand Margaret my gun and tell her to pull my fish in. She and her boss at Dow were fishing on the stern standing on the platform. I handed her the gun and it almost jerked her in the water, her boss Gary grabbed her and the gun and saved the day. He then pulled my cobia in. I felt like an idiot, and she was upset. She could have been injured by my foolish act..

Freeport Tx The Pratt's

The first time I used the No Ka Oi Glenda and I were out in our Seacraft with our son in law Andrew. The vis was very poor that day. I made out a shape, it was either a shark or a cobia, so I shot. All hell broke loose. I had a big fish and it appeared unhurt. After a struggle I handed my gun to Andy and asked for another gun. Every time I tried to approach the fish to put another spear in it, it would freak out and I couldn't place a shot. It stayed calm as long as I let it stay 20 ft. down. After several tries I climbed in the boat and Andy and I pulled it into the boat. It went crazy, we just got out of the way. It ended up breaking my old teak splash board. At the dock it weighed 74 lbs.

On another trip in Bob's boat I shot a cobe in the 50 lb range and was teasing Bob about, this is how it's done, when he shot one in the 60 lb range. Clue, never tease and gloat. We had a blast in that old boat fishing and diving. I got their daughter Taylor in the water one day with dolphins. She was hesitant, but ended up enjoying the experience. As I aged I would be exhausted as we reached the dock. I had certain tasks I forced myself to do, but if allowed, I would just go for beer. I always loved beer at day's end, even though it made my exhaustion worse.

Over the years I shot so many cobia they became my least favorite fish. On some I had to pull my gun back to let them get far enough away to shoot. My joy will always be snapper, red and gray. Stalking and placing my shot with my new open muzzle 2 sling, 1100 Rob Allen Tuna gun. Rob and Jeremy copied Riffe in letting the shooting line hold the shaft in place. My technique on king mackerel was to ascend away from the rig where they circle. Most macks I have shot were shot on the way up. To me the roller guns are a fad, I would shoot just about anything with my RA except tuna. Thinking back though, those brittle shafts would not have worked in the caves of Anegada, I would have needed SS.

WE ALMOST LOSE THE 23 SEACRAFT

The last local fun Fishing Fiesta we ever did was almost a disaster. Glenda and I had our son in law Andrew Meraz with us. There were menacing black clouds to the South, so we headed East. We arrived at the 20 mile rigs with the weather looking kinda scary. After attaching our boat hook to the rig, I quickly grabbed my gear and piled in hoping for a quick shot at a big snapper.I was on my 3rd dive when I noticed the waves were getting huge. I looked towards the boat and it had straightened the aluminum hook and was drifting away. Then the storm hit and I began to realize we were in trouble. I saw a charter captain pass by heading for the jetties. I swam as hard as I could to catch the boat. Glenda had pulled out the life jackets and was frantically trying to drive the boat (full throttle in neutral) I scrambled aboard and got to the controls, (we were sideways to huge seas), pulled back the throttle and put it in gear and turned towards the jetties. Now we were in blinding rain and huge seas.

I love my Seacraft and I knew we would make it home, we had a new 200HP 4 stroke Suzuki and I had excellent control of the boat. I had to put my mask on to be able to see, this was about the worst sea I'd been in in this boat. I was scared we might hit the jetties in the blinding rain, and still remember seeing the boulders only 30 or 40 feet to my right as we entered the harbor.

I'll say it again, Glenda is a saint, married to a crazy spearfisherman. Over time they took spearfishing out of the July 4th fun tournament due to lack of participation. They still show my 321 LB jewfish every year in their promo booklets.

2005 KATRINA

Just like Hurricane Hugo had made John McCallum rich, Hurricane Katrina made Louis Schaeffer rich. Luis had started another commercial diving company near New Orleans, he was in the right place at the right time. The oil companies in desperate need started handing his company major contracts. His company grew by leaps and bounds. Louis went public with an IPO. I googled him ringing the opening bell on Wall Street. After a few short years, Louis retired with many millions. The company went bankrupt shortly after. Don't ask me how this works, I've never had real money, but I knew not to buy stock in anything Louis. Louis came back to Freeport and bought the Big E a very large charter boat and the lovely old building where my daughter Kim was married. Poor white trash boy who always needed help and money made good, and came back to show off. I got his phone number from the boat captain, but he would not answer or call me back. Maybe he thought I wanted back wages.

Years earlier Louis was with us on a crazy trip to the 538 rig. Thomas McDonald had a steel hull boat with a big diesel, he called it the Turtle. Tom put together a trip with his son Craig, Jim Rio, Louis and myself. The seas were up, we shouldn't have gone. We were pounding right into the seas. Craig got sea sick and was begging his dad to turn around, Tom kept saying just a little further and we'll be there. Louis had also been known to get sea sick. We pounded and pounded and the boat began to wallow strangely. Jim and I lifted the deck hatch and saw that we were sinking. Jim and I grabbed buckets and jumped down into the spacious bilge area and started bailing furiously. We got the bilge pump going and finally made it to the rig. Tom McDonal was a big strong tough man. Louis with one leg was not too anxious to get in the water and probably seasick. I'm guessing 4 to 5 ft seas. So Tom started putting his gear together for him and helping him get ready. When Louis was finally ready, we picked him up and threw him over the side. We never worried about Louis drowning, he was a waterman. Jim Rio and I would dive in almost any condition, we both ended up getting nice big red snappers on scuba, I don't think anyone else got anything. I was proud of the fact that years before when salvaging the Middleground, Louis had told Gloria that I was one of the best divers in the gulf. I told this story years later as part of Tom's eulogy, Tom passed at 81.

TDF AND KURT BICKEL

Kurt Bickel was on the freedive list and the editor of Spearing Mag. I found out we were both cyclists and he encouraged me to join him on a bicycle tour that would follow the Tour De France from Marseille to Paris. I did and we did. I think it was 2004, that would make me 68. I thought I was quite fit, but I ended up riding with the slower group. A few of us wanted to try the Bouillabaisse soup that Marseille was famous for. We paid $50 for some fishy soup. Dumb tourists. It was a great experience.

Natural fountains you could drink out of in the public square of small towns, beautiful fields of lavender and hotter than hell. I came down a long slope one day to a boat ramp that ran into a beautiful lake. I was so hot I rode right into the lake. Lance Armstrong was my hero then, the world had not discovered what a worm he was yet, although the French were highly suspicious. Betsy Andrau, whose tenacity helped finally expose him, is my Facebook friend today. No one really wanted to expose him, they were all making too much money off of his fame.

Kurt had a bad experience at the midnight lump in Venice La. when a mako shark hit him and drove him up and out of the water. He eventually left spearfishing for the cycling business and was very competitive in senior cycling events.

I entered the Houston Senior Olympics cycling 2 years in a row and got one 3rd place. The 2nd year there was a terrible crash. I was not in that group, so we kept racing, passing the ambulance as they were loading him. He died. That was enough for me, who cares which old guy wins a bike race.

Over the years we vacationed twice in Alaska and once in Hawaii for the 60th anniversary of Pearl Harbor. I was a 5 year old Navy kid PH survivor. We met Dorinda Nicholoson who was my neighbor. I was 5, she was 8. She wrote a Book Pearl Harbor Child because she was told (like me and my sister) she was not a PH survivor when we tried to attend a survivors annual reunion. Anyone wanting to know my take on PH can contact me on FB, I will email you the annual statement I post every year on Dec 7th. PH was bait, to encourage an attack, enrage a pacifist public (after WW1) and save Europe.

I did an AARP triathlon while in Oahu. I swam and cycled but didn't run. I found a guy to run for me. Turns out he was a world famous runner Tom Knoll, an original Ironman. He had also done the run across America. We would have placed 3rd in our age group but I missed a turn on my bike. Tom passed in 2018.

TWO YEARS IN A 5TH WHEEL

Having been in our Bridge Harbor House for 10 years and evacuating 4 times we were ready to move on. Dow had laid me off when they bought Union Carbide because I was a contract engineer, and they needed to place the new Carbide engineers. We sold the Bridge Harbor house, bought a 3 axle 5th wheel and a 350 dually Ford Diesel truck and took off to see mostly Western America.

After 2 years of RV parks we were burnt out and yearning for a home again. I emailed my boss at Dow and went back to work in Houston. We sold the 5th wheel and bought a house in Angleton Tx. I could go 50 miles to work in Houston one way or 15 miles to Freeport the other way.

MOTORCYCLE

Most guys sooner or later want a motorcycle. Always money conscious, I bought a used Kawasaki. It was a nice machine. Riding locally gets boring as I rode alone. We still had the truck so we bought a Lance truck camper and a trailer for the MC. For our next vacation we would visit friends in Creed Co., Glenda's sister in Oregon and Kathy McCallum in Boise Idaho. It was a fun trip. I rode through the mountains (almost hitting a deer in Colorado) but put it in the trailer for long stretches of desert or plains. Riding down California coastal Hwy 1 was a beautiful experience. I would come around some curves and let out a scream at the beauty. We stopped to see Brandi Easter along the coast.

Before reaching San Francisco we came to a very steep switchback, Glenda following in the truck. I was gazing at the beautiful cliffs and ocean when I realized I needed first gear, too late. I fell off the back going head over heels bonk bonk bonk (my helmet) Dazed and confused I finally stopped and got up and walked back to the MC. I sat down holding my elbow in pain. People kept stopping wanting to take me to the ER. No, I'm fine, I replied. A logging trucker screeched to a halt and jumped out to help me. He looked like an all American athlete. He grabbed my bike and held it up for me. I got the engine going and thanked him profusely. Glenda had found a flat spot to pull off up ahead. I rode up and put the bike away. My riding was finished. The dive back to Texas seemed never ending. I sold the MC, and later the truck and trailer.

2010 - BILL CRAWFORD

Bill and I had communicated over the years on guns and fish. He liked to fight fish with his reel line. I cautioned him that in deep water shooting very large fish (big AJ's, tuna) is safer with a float. I got news that Bill had drowned. Out spearing with Chad Morris, Bill's gun had floated up and he was missing. Chad got a group of local scuba divers together. Next morning they circled the rig in a pattern and found Bill. Leaving a wife and 2 daughters, this was extremely sad.

My theory is that if he was using a closed muzzle Rob Allen, his shooting line may have wedged with a rubber in the muzzle, not having access to his reel line he tried to horse the fish up to the surface, realizing he couldn't make it, rather than lose his gun he cut his shooting line and then blacked out. Another possibility is a fish wrapped his line in the rig and he cut it loose to save his gun.

A few years earlier I was doing a job for Dow in Baton Rouge. I called him to try and get together. We met at Chimes, right near the LSU campus. He was a very pleasant guy and we had a great conversation. I am sorry I never got to dive with him. He was very popular with all the Cajun divers and famous for spearing wahoo in his small aluminum boat and a hybrid snapper he had shot that looked like a gray, but was way larger than any gray (mangrove) on record.

THE CAJUN CREWE

Joe Wegmann organized a memorial tournament in Bill's honor and I wanted to attend. By now I was in my mid 70's, slowing down but still loved to shoot. Joe found me a boat ride with Jimmy Richard. Jimmy was approaching 60 but was also a cyclist and a fit freediver. Jimmy had a smaller contender and space was at a minimum. It was a rough day offshore. John

Delao and Scott Cameron were on board. I got a nice cobia and after jumping about 5 rigs I was exhausted. These guys must have jumped in at 20 rigs. They shot a ton of gray snapper and some cobia.. I begged off the 2nd day and Glenda and I went for a tour of the Avery Island plantation where Tabasco hot sauce is made. A very worthwhile tour. What goes into their products made me a fan for life.

Tracy Palmisano won the first year. He is an interesting guy who owns the Boat Yard in Nola.

He won a custom speargun made by Chad Morris who was with Bill on that fateful day. Tracy was proud to show me he already had 2 Daryl Wong guns. He is an avid freediver. He was amazed I knew Louis Shaeffer and had helped Louis salvage the Middleground. He knew Louis quit well. I gave my cobia to the guys but I had told them I loved gray snapper. While I was talking to Tracy, they got with Glenda and filled our ice chest with gray snapper filets. A wonderful act of friendship. Friends for life.

My second trip to the memorial tournament, which Joe calls "The Louisiana Freediving Cham

pionship " was the most fun. Joe got me a boat ride with Vincent LeBlanc. Vinny had a cattamaran with 2 outboards, it was a great dive boat. We were on our way to visit my oldest son in Alabama. On board that day were Vinnie, Abbey Gail (now Woodard), John Delao, Jimmy Richard and Scott Cameron..

I was having a hard time in rough seas getting to the rig and back in the boat. Further offshore I fanally let everyone get in the water first and out of my way. I worked my way to the bow and fell in, then swam a short distance and grabbed the tie up rope. Everyone was close to the rig while I was out by the boat. My technique as I aged was to find something to hold onto while I breathed up and rested my legs. Resting your legs is critical to comfortable dive time as you age. I made my dive well away from the rig, broke into clear water and saw a large school of red snapper below. As I descended one large snapper rose up for a look, just as it turned to start back down I placed my shot just behind the gills. I started up, releasing line from my reel. I was well away from the rig and had kevlar line, so I wasn't concerned the fish would find its way into the barnacle encrusted pilings.. I was using my RA Tuna gun so I was pretty confident that my shot had gone through and my flopper was open on the shaft tip. I swam to the boat and handed someone my gun and started pulling my fish up. They were blown away. They asked me if I had ever shot a snapper that big before. I said well in my scuba days salvaging the Middleground I had shot one like that. I managed to get 2 more medium snapper at other rigs before we started in.

Abbey is very athletic and aggressive. She landed a big cobia and some snapper. Her mask messed up and she asked to borrow mine. I was pooped out. I use a Cressie Big Eye with a Scubapro shotgun snorkel. She was impressed with the snorkel, she said the volume and easy clear valve gave her an extra 15 seconds dive time. I always kept my snorkel in my mouth. In fact I would catch myself in the boat with my mask and snorkel still in place, they were so comfortable. I never understood why anyone would want to clear a straight snorkel and I didn't want to spit my snorkel out and then have to put it back. Later I lectured Sheri Daye on why she should try this snorkel, she is so nice she didn't tell me where I should stick my snorkel. I tagged Abbey and she added agreement.(on Facebook)

The most exciting thing on that trip was Abbie's bikini. Her husband today Travis, is also a spearo, they make a great couple.

At the dock Abbie was laying her snapper on mine to see if hers might be bigger, the guys teased her unmercifully telling he mine was much bigger than hers.

I was a zombie. It always hits me at the dock, extreme fatigue. I got my big snapper and found a cleaning table. The guys were upset I wouldn't save my fish for the weigh in, Jimmy Richard said I was cleaning the first place snapper. I really didn't care about winning in spearfishing any more, it was just the companionship and the joy of mother ocean that remained an addiction. John Delao brought me another snapper I had forgotten about. As I finished in the heat I was close to collapsing. Later at my son's house 5 of us ate off of that snapper for 3 days. My guess would be in the plus or minus 30 lb area. Jimmy Richard had the picture, but when he left FB it was gone, I haven't seen it since.

I think Blake Bidwell won that year. He was diving with Lance Williams. Blake posts lots of fish video's today on FaceBook. He is a great athlete.

The cajun crew

Our last memorial trip to Port Fourchon. I'm not sure what year. But it was the worst trip. Joe got me a boat ride with the Tuna Captain in a large Freeman. Lovely boat. The water was brown and rough and the current was ripping. On a good day this area is one of the fishiest in the world, on a bad day stay home (unless you're young) I should have asked the captain to drop me off upstream of the rigs and let the current carry me to the rig like we did in the scuba days. I never shot my gun. Not sure how old I was that year, maybe 79.

There were 3 young guys on board and one especially wouldn't give up. Brandon Hendrickson jumped rig after rig after rig. I later read an article about extended breath holding and brain damage that featured Brandon. I don't think any conclusions were reached. When the captain asked if anyone wanted to hit another rig Brandon always said yes. I don't remember them getting much.

LA PAZ AND TIM HATLER

Glenda and I had heard so much about La Paz we had to try it. We visited Palapas Ventana twice and the Cortez Club and hotel once. Tim Hatler and Jimena were gracious hosts at Palapas Ventana. Our guides with Tim were Yoni Lucero and Gilberto Olachea. I wanted to get a wahoo but never did. Tim had a target set up to see how your guns shoot. I love the Riffe family but horizontally the #2 standard shoots high and the No ka oi shoots low. I should have spent the money for a euro design gun. I hadn't bought my RA tuna gun yet.

On 2 occasions Glenda could have shot one, once while I was reloading she tapped me on the shoulder and I looked up at one very close to me watching me reload. Of course as soon as you are ready all you see is a tail leaving. Another time I was bored and hanging out at 30 ft by my flasher and one swam right up to Glenda on the surface. They know who has the gun. This type of hunting is not for me. People who hunt oil rigs are spoiled, they just keep moving rig to rig till they find fish. Rig removal has now even made that difficult. (except in Fourchon La)

I had gotten to know Ron Mullins on the Freedive list and we stayed in touch. He had built a compound in La Paz and he agreed to come over and give us a tour of La Paz. We ended up spending the night with his neighbor and we met his wife Michelle.

So the following year we booked a week at the Cortez Club. Their guide was an Italian named Toto Volpe. He had been to the nationals more than once and was a very good spearo. Isla Espiritu Santo is absolutely beautiful. Small parts of it were open to spearing. Toto took us there. His technique was to breath up, go 60 ft. to the bottom in his camo suit, become a rock and wait for a fish to move.. He was very good at it and got some nice fish. I could go to the bottom and hang out for a short time, then I had to come up for air plus I did not have camo. He wanted us to go again the next day and he would train me. I told him I wasn't good enough to shoot fish the way he did. Instead we got on a tour boat that took us all over Espititu Santo. We snorkeled in a protected area and saw lots of beautiful fish. The little sea lions were fun to play with, but there were also monsters that were aggressive and scary. A girl there told us she had been bitten by one.

Toto told us that locals were going out at night with hookah rigs and plundering the area. He said if I had been there a few years earlier I would have done well, even with my diminished bottom times. That reminded me of Anegada where they would scrape the eggs off of lobster to cook for tourists and how they wiped the conch out. When we first got to Anegada in 1976 the South East end was covered with queen conch. The last time I was there, there were mountains of conch shells and empty sand. Lowell Wheatley's brother Capt. Blondie bragged to me he was going to have 100 fish pots some day.

We met Manuel Gonzalez who wanted to give us free diving lessons. He told us he was the free diving champion of Mexico. He was a nice young man. I see him on FB today, recently into technical diving.

They had a monster margarita there at the bar that was expensive. I complained about Glenda getting one, so she drank two. I had to lead her to the room and pour her in bed, drunk as a skunk.

Somewhere along the way I had a cochlear implant. I wore 2 hearing aids for quite a while. My right ear got so bad I needed the cochlear. I had tinnitus for many many years from firecrackers and guns when I was a dumb kid. Traveling to BCS (Baja California Sur) I carried my Senheiser neck loop with an FM transmitter. I plug it in a TV, computer or phone and put my gadgets on T coil and the sound goes right into my ear.

NEWS OF 4 TRAGIC DEATHS IN THE VIRGIN ISLANDS

John Stuart Jervis had flown in the RAF. He gave me a bi annual flight review one year and made me fly in clouds. Not fun. He also taught Glenda ground school. He was an avid balloonist. I read where he had been shot down for ballooning over Belarus without permission.

Mark Marin, one of the Marin brothers in St Thomas who I dove with often, fell down a flight of stairs and died. He was the headmaster of Antilles school.

Ariel Bristol, one of the best pilots in the VI. Trained and soloed Glenda in our plane. Died hauling cargo in very bad weather.

Lowell Wheatly. Owner of Anegada Reef Hotel. I had gotten to know Lowell very well. Attempting to make a BBQ pit from a 55 gal. drum,blew himself up and died. Vivian and the kids Lowrance and Lorraine run it today. Their rum smoothies are famous.

PANAMA

Jim and Robin Rio moved to Boquete Panama. After researching the area they bought a home and settled in. When it was time for our next vacation we visited them for a week. We fell in love with the area. Boquete is high enough for a wonderful year round cool temperature and had become a retirement mecca. We decided, this is it, we'll retire here too.

We found a very nice condo right in town at Valle Escondido resort. It was a large resort with a 9 hole golf course, indoor pool and spa, amphi-theater for concerts, and a very nice restaurant with an outdoor bar. We were moving to paradise.

We lasted 9 months. The Rio's lasted 2 years. (they had a house to sell) I had read all the propaganda about the wonderful spearfishing in Panama and Costa Rica, I was ready for it, in my 70's but still good at 60 ft. if I could rest my legs and breath up. I would train in swimming pools swimming UW laps.

Jim had made the acquaintence of a local legend Bill Fitz. A former photographer for National Geographic, he had settled in Boquete and had a flourishing business selling tropical plants all over the world. He explored the Panama jungles alone, finding flowers and plants to bring back to his "flower ranch" His business flourished.

Bill had a house down on the coast where he kept his boat. It was a 3 day trip to go diving. One day to drive to the coast and get everything ready, one day to dive and shoot because you had to go out and return based on the tides and one day to clean up and drive home. Jim and Bill got some nice snapper on most trips using scuba at 100 ft plus. It was supposed to be illegal but there wasn't any enforcement. When Glenda and I were invited I didn't get squat. Good fish were not available at my depths of freediving. I could shoot oceanic triggerfish if I wanted one, but that was about it. Not like the rigs where you dive down 30 to 60 ft and shoot nice size snapper or have a cobia come by you at 10 ft.

We bought the Panama standard, a 4 wheel drive Toyota Hilux diesel truck and explored and enjoyed the area for a while. I loved that truck. One popular trip was to drive to the east coast and visit Bocas Del Toro. Jim and Robin took us the first time. It is a beaches and party area. You park in a protected compound and are ferried over. I kept trying to find out about spear fishing but only got vague answers. Later Glenda and I went alone and I found a guy who would take me spearfishing. Sportsfishing Bocas del Toro. Can't recall the guy's name, he was a Hawaiin and he was willing to take us all over the place to spearfish where there are no fish to spear.

Glenda brought Kim and our grandkids down and we did white water rafting and the zipline. The zipline scared the hell out of me. I couldn't hear the safety talk and when I approached a tree at 50 mph I freaked, not knowing they were going to brake me in the last few seconds.

We made a trip in our Hilux to the peninsula that divides Panama and Costa Rica. Again no maps. Puerto Armuelles is impossible to get through without experience. I flagged a taxi and we followed him through the city to the exit road to Hooked On Panama and Limones. We had a pleasant surprise at Hooked on Panama, the owners were from Freeport Texas and they knew Louis Shaeffer and many others we knew.

Next came Bob and Margaret Pratt, we did Bocas again but stayed on a remote island rental in a jungle. At supper I noticed a python type snake moving along the rafters. To drive Panama today I recommend Google Earth.

TIM HATLER'S TUNA HUNT

Tim put together a tuna hunt at Hooked on Panama. I signed on and prepared to drive alone with my birthday cake. There were about 18 spearo's that would be divided between 3 boats. The Hooked On Panama family had some great boats and captains. Their parents had been in the boat building business, in fact their father had built my first boat, the Hurricane I used in the 60's.. Wow, small world. At dinner I broke out my birthday cake with a small piece for all. The guys were blown away that I wanted to hunt tuna at 78. Other spearo's there wanted to ignore their own birthdays in honor of mine.

Tim had a huge gun he wanted me to use. I had my No ka oi. Tim had always done everything he could to help me succeed. We ran the hell out of those boats for 5 days. Chasing tuna boils sometimes right in the middle of commercial fishing boats. The water was cloudy and I only saw vague shapes over and over. Porpoise were mixed in with tuna, we would run and run and dive and dive. Once I made an extra deep dive, it was creepy, dark and empty. It became obvious you needed a quick shot near the surface. 18 divers, 5 days and we only got 3 or 4 tuna. We ran through some terrible rain storms, with our captains on top in the open while we huddled below.

Actually, like most places in the world, it turns out the only real place to hunt is 60 miles out, Isla Montuosa. We managed one day there. It is right next to Hannibal Banks and has not only tuna but huge snapper and cubera. I'm getting used to coming home empty handed when I travel. Heck, just give me my oil rigs.

I met a guy named Ted Tennis, he is very good and he got one of the tuna and gave me some. He sold real estate in Pedasi and invited me to visit. He knew Bill Delabar who had been the captain on my first trip to Venice La. Bill was in Panama now running a yacht for the ex president of Panama.

Bill Fitz hunted tuna every season. His boat is too small to carry more than one passenger with all the supplies they needed plus extra fuel. He and Jim Rio went to Montuosa to hunt tuna. Jim said he spent the most miserable night of his life trying to sleep on the island. Bugs. Bugs will eat you alive there. Like Tim's group they chased tuna boils and jumped in time after time with no results. They made more trips and Bill got his tuna and finally Jim got his. He said one almost ran over him after he had missed an earlier shot, these tuna are in a frenzy as they rocket around while feeding. A person that can find traveling tuna swimming calmly in clear water has a much easier shot. We were trying to shoot in the middle of a boil. BTW, 6 guns, 6 floats and float lines in a boat is one helluva mess to keep straight.

THE GOOD, THE BAD AND THE UGLY

Glenda went home a lot so she didn't have to leave to get her passport stamped when I did. In Panama you will always be a foreigner. You have to leave every 3 months and get your passport stamped. I would drive over to Costa Rica and spend the night and drive back.

However, I noticed connected wealthy people ignored all that. The Indians do all the work, live in poverty and the men get drunk and lay in the streets every Sunday morning. If not for the Canal Panama would be as poor as Costa Rica. Resorts shield tourists from the poverty. Pedasi is on the coast, they don't see the poverty that is in the mountains where coffee beans are harvested. Pedasi is booming, Ted Tennis was doing well in real estate. If I returned I would choose Pedasi.

To get into your bank you have to remove your hat and not wear sunglasses. The outside guard approves you and the inside guard lets you in.

Military on motorcycles patrol, if they stop you, you better have your passport. Socialized medicine is ugly. However the country is beautiful and the climate in Boquete wonderful.

I was in the Boquete town square one day when I noticed an extremely tall girl just outside a hostel, she had to be at least 6 ft 8 or 10 inches. Puffing a cigarette, her image sticks in my mind today. A few weeks later the news was full of two Dutch girls who went hiking to Bocas del Toro and went missing. One of them was very tall. The drive is lengthy, down, over and up the coast, but the Indians had a trail blazed straight across the mountains. A long search ensued with no results. Finally the Indians got involved. They only found a shoe with a foot in it and bits of clothing. Then the rumors started. Organ harvesting? Wild animals? Not sure if any conclusions were drawn. Such a sad story.

PEDASI

Glenda and I made a trip to Pedasi. Ted Tennis is a fixture in town with an eatery across from his office where he is part of the band with his guitar. Someone was having a party the night we arrived and they invited us. Wow, just like Texas very friendly people, we fit right in and felt at home. It was a younger crowd. Boquete was mostly old people that sat around talking, drinking and smoking.

I made a dive trip with Ted. Nothing in Panama is easy. Remote boat launch down a steep hill to a beach, out 2 miles to some rocks to dive. Vis was the usual dim 8 to 10 ft. I had gotten used to. I kept diving as deep as I could but never saw anything. Ted dove deeper and finally got a snapper. A small whale shark came by and Ted got a vague dim picture.

Glenda and I spent a day snorkeling Isla Iguana,the local park and sanctuary.. We rented a panga for the day. We had about 30 ft of visibility and we got our hopes up, there should be nice fish here. The guide told me to bring my spear gun and he would show us where I could hunt. We snorkeled for about 4 hours on the protected side and never saw a grouper or snapper, or any real food fish. The guide took us around the island to the back side for some spear fishing. Nothing.

I worried about my speargun being in the park. There were military types hanging around. They were friends of the guide, no problem. I began to see why there were no food fish in the park. It is Iguana Isle. and they are everywhere. We became tourists and tried to just enjoy the trip, the water and the scenery.

Back on the beach I helped the locals by using our Hilux 4WD to pull some pangas around on the beach to get them where they wanted them based on the tides. Everything is hard diving Panama.

BORED

Back in Boquete we got into a rut. Spearfishing sucked. Every trip was a major undertaking. We didn't get invited with Fitz that often, and that was a 3 day trip with poor results for an old freediver. We did buy wonderful fish from the commercial fisherman in town, brought up from the coast. Panama does have good spearfishing for illegal scuba divers and those that can make it to Montuosa. We were making trips to David to shop for stuff we didn't need. Socialism is interesting. In David there were huge stores full of stuff and people to sell it and no customers.

I tried to sell my Seacraft when we arrived in Panama, positive I would die there. The best offer I had was 12K. Hell I had just spent 14k on a new 4 stroke Suzuki, plus a new custom T top and leaning post. Everyone now seemed to want a Contender, not a 1982 Seacraft. I contacted my oldest son in Alabama and told him he could have it for 10K and could take his time paying. So he and a friend drove to Angleton and hauled it back to Elberta Alabama where it sits today in his backyard.

I announced to Glenda "we're going home" she was shocked, but more than ready. We had done Panama.

While waiting for our flight home in Panama City we became tourists. The fish market there is mind boggling with huge amounts of all types of food fish. We visited old Panama churches and other tour sites. Then we each managed trips to the ER with food poisoning. Glenda first. A cab rushed us to the best hospital where they weren't sure they wanted her. I would need $400 cash for admission. I pulled out a wad of hundreds and handed them 4 and they were happy. A day later it's me, rush to the state hospital, for fifty bucks, same treatment but poor decor and housekeeping.

COCHLEAR IMPLANT AND HEART ATTACK

Arriving back in Angleton Tx my right ear was about gone. I was approved for a cochlear implant but only after I convinced a cardiologist that my bradycardia was normal because I was a freediver. Amazing technology. The sound was weird at first but I soon learned to love it. I'm happy my left ear only needs an aid. When they put the electrode into your cochlear, it kills your ear. I now hear on the right side through my skull. An MRI tells them exactly where to drill the hole to install the electrode. The implant has a magnet and coil plus an amplifier. The external magnet attaches and is surrounded by a coil which vibrates like a radio speaker cone. People who have nerve damage from the cochlear to the brain can't qualify. Did you know you see and hear with your brain?

We had just arrived home when a repo man at the wrong house was pounding on our door wanting money for the motorcycle he said we had bought. I heard screaming at the front door. Glenda is very high strung. They kept screaming at each other. I came and looked through our small glass opening and saw a muscled up guy full of tats playing TV repo man. Glenda had already called the police. We knew one of the cops. They told the guy to get lost and that he would be arrested if he came back on our property. Glenda stepped outside and asked him for an apology. He told her he was free, white and he didn't owe her anything. She does not handle stress well. She felt ill for an hour or so and ended up in the ER. They rushed her to Methodist. If I had answered the door I would have simply told the idiot to get lost, he had the wrong house. Methodist Hospital saved her and now she has a stent. She obtained a lawyer and some day it may end up in court.

LAST BOAT

Bob and Margaret Pratt had visited us in Panama. They were both retired now and building a new home on 20 acres. The Kate C just sat and deteriorated. I had hoped that I would be diving on that boat till I croaked. Not gonna happen.

We found a small lake type glass boat with a 4 stroke 50 HP Honda. It was only a few inches above the water with an elevated bow. I wouldn't need a ladder to get in. And we could fish inshore in shallow water. It ran great and the trailer looked new.

We used this boat for about 3 years. One summer we caught a flat day. My favorite rig was 13 miles south of the Freeport jetties. We headed for the whacker rig. There were only a couple of boats in the area. I got in and started snorkeling around, not expecting much. On a shallow dive I spotted a sow near the bottom at 60 ft. I did a long breath up and dove, I was able to place a perfect shot. I was overjoyed. I hadn't shot a red snapper in a long time. I continued hunting and about 30 minutes later shot a bigger one. 14 and 18 pounders, I was a happy man. We had run into a friend before we left and he said "are you going offshore in that" we said sure it's flat out there. You're crazy,he replied. While cleaning the fish I wondered if these were my last snapper. I was approaching 80. We made other trips in the little boat trolling for king mackerel with some success. Glenda had a bad fall on the dock and she saw me make a mistake running the engine out of the water, she said we're selling the boat. Boatless again. Guess who was always in charge. I never had any success arguing with her, and she never told me I couldn't go diving or fishing.

Last 2 red snapper

KEITH LOVE

I had heard about Keith Love chartering spearfishing trips. I chartered a trip with Keith. We had a pretty good day. He had scuba and free divers. The scuba guys were impressed that an old man "without tanks" had gotten some nice red snapper. Daniel Adams remains FB friends today. Also on the trip were the Padua brothers Alex and Tony. They are good free divers. I got upset at Keith and Alex for shooting a huge stingray. Why kill this fish? I thought we were through for the day and I put my dry clothes on, But Keith stopped at another rig and we ended up next to a rig rope hanging down to the water.. Keith got in last and shot a nice cobia. After he boated it, he got back in and swam to the rope next to the boat, held on to the rope and breathed up and dove. Came up with a nice red snapper, then he did it again and again. I was pissed at myself for putting my gear away, that's my best game, holding a rope, resting my legs and diving. Damn, I should have done that and got more snapper.

I backed out of a later trip because Keith had too many people on board. Brandon Hendrickson among them.

On the next charter I booked I met Willie Powers and Patrick Gaudi. The thing about chartering with Keith is the interesting people you meet. Keith brought along a bikini. If I had known how much room we were going to have I could have brought Glenda. Keith is pretty easy going and loves to shoot and have fun. He comes from a wealthy family and tends to attract hopeful bikinis. We went out to the little Bucs. These rigs are in 80 ft of water. We found snapper and everyone got busy. The snapper were holding at about 60 ft. I could barely reach them and took one bad shot. As the rest of the guys were boating fish I was unable to get down far enough for another shot. I had been diving and harvesting snapper at these rigs for many years. The limit is 2 and Willie had his 2 when I asked him to shoot 2 for me. He did. He is a good and generous spearo and person. He and Patrick are also great personalities and fun to dive with. Willie had some beer and I love beer at the end of the day. Glenda met us at the dock and I was turning into a zombie again. The diving, trip and the beer exhausted me. This was the first clue (to me) that I needed a pacemaker. All of us stayed connected on FaceBook. Patrick is from a wealthy family in Mexico City. (I assumed, because all he seems to do is travel the world, minding their businesses and have fun)

Keith Love trip

HYPOTHERMIA

Keith does winter trips for wahoo 160 miles offshore, so I made another desperate attempt for a wahoo. I bought a measured wetsuit online from Bare. I hated it. Putting it on my old bones was very difficult.

The Padua brothers Tony and Alex were on board along with Pablo Perez. Keith usually spent the night at a rig in the area where he collected fish for his aquarium. Everyone got in the water. Big mistake for me. I started getting cold. Pablo shot a large AJ that damaged his large gun and float line before he lost it. It was a long cold night for me. I couldn't sleep or really get warm.

I think I was 79 or 80.

Next morning Keith ran to the hoo area, we suited up and got ready. My adrenaline was pumping. Keith had a deckhand to run the boat while we all hunted. I was in the water right away. Another mistake. We didn't find fish for a while. So I was in the water much longer than need be. Finally we saw hoo. I followed trying to work my way closer for a dive. I hyperventilated getting ready for a dive (this is totally different from rig diving and resting your legs while holding on to a rope or piling) I dove and kicked as hard as I could, trying for a closer shot, I finally shot and missed. I barely had the strength to reload my gun. The boat came back to retrieve me. Keith's customers are young and strong and don't need a boarding ladder. You have to pull yourself up on the transom next to the outboards as best you can. I handed my gun to the deckhand and attempted to climb aboard. I was so weak I struggled as hard as I could and finally made it with the deckhand wanting to assist me. I went into the zombie state. I was finished. Keith wanted me to put a wetsuit top over mine to warm me up. I tried, but was too weak to pull it on. I peeled my wetsuit off with great effort, put on as many warm clothes as I could and found a place to cover up and sit down. I should have brought more warm clothes.

Pablo asked if he could borrow my float and float line. Sure, I said good luck. He had a second smaller gun, so he rigged up, suited up and was ready to go. They finally got into the fish and it was hot and heavy with excitement for an hour, running here and there, gaffing fish and picking up divers. That's when Pablo got in, perfect timing. The Padua brothers each got hoo along with Keith. A couple were lost to the ever present sharks. Finally Pablo got his fish. It was chaos and messy with fish, blood, guns and float lines everywhere.

I maintained my zombie state for the long ride home. In fact I was wondering what it feels like to die with hypothermia. I tried and tried to warm up. I don't think I ever got warm until I got into a warm bath at home. This is it for me. I gave that wetsuit to my grandson Nathan Meraz.

Back on the dock Pablo was elated. They were all happy, but Pablo was the happiest. He shared a nice chunk of wahoo with me. We love wahoo steak at my house. The trip cost $800 each so this decentivised the Pagua brothers from sharing. I maintain a warm relationship on Facebook with Pablo today. He has a wonderful family.

Looking back, I'm sure I almost died on that trip. If not for staying fit at the gym, UW pool laps and cycling I would be gone.

THE BLUE WILD AND THE BAHAMAS

Glenda and I decided to attend the Blue Wild. I talked my oldest son Stan into joining us with my granddaughter Trinity. The years have run together for me, it could have been 2016 or 17. I considered chartering a Bahamas trip with Cameron Kirkonnell, but he kept adding people, airplanes and money until I backed out. I couldn't get a firm price. The Blue Wild was a fun trip. I got to meet some Facebook friends plus Stan Waterman. Daryl Wong and Sheri Daye were very busy but said hello. I saw Jay Riffe and Jill again. Just a fun trip, but a long drive from Texas.

Willie Powers and family go to Green Turtle Cay every year in the Bahamas. We talked it over and invited Bob and Margaret Pratt to go with us. We rented a VRBO and reserved a boat from Sunset Marine. We ended up going 2 years in a row.

The first year we arrived 2 days before Bob and Margaret. We got our boat with a guide and went hunting. I had a pole spear which I haven't used in 30 years. After an hour of searching I spied a nice Nassau grouper, took a breath and dove. It was up against the reef in plain sight. I couldn't believe I could get so close, it must have thought it was camoflauged. Instead of pointing the pole like a speargun, I held it under me and hit the reef above it. I felt like an idiot. That was the biggest grouper I would see both years. Good gawd, I am a proponent of pointing at the spot you want to hit.

KNOCKING ON HEAVENS DOOR

I was having trouble seeing detail on the bottom, so I started cruising 20 ft down for a clear view of everything. I was making an unbelievably long cruise, feeling no need to breathe when suddenly I had to vomit. I headed up as fast as I could in panic. At the surface I went into rapid uncontrolled monster hyperventilation, and hollered for the boat. At the boat I struggled on board and could not stop hyperventilating. I started crying and Glenda was hugging me, we all thought I was dying. No paper bag around, but after what seemed like an eternity it eased off and I calmed down. (more about this at the cardiologist much later)

Knock knock knocking on heaven's door, I can't shoot fish, anymore, getting dark to dark to see, I can't shoot fish, anymore. I had definitely knocked on heaven's door, but it didn't open.

Next day on the beach where they feed stingrays (small sharks join in) Willie's father in law told me I should have cupped my hands over my mouth to ease my condition since we didn't have a paper bag. Willie's father in law had gained some notoriety in a previous year by being hit in the face by a shark while surfacing holding a snapper in front of his face. The reef right outside the feeding area holds some nice fish and some nasty little sharks who want to be fed.

Green Turtle Cay is a neat place. Provincetown was settled by loyalists to the king during the American revolution. Tourists rent golf carts to get around. There are resorts or you can eat off of local vendors. We explored the island and snorkeled everywhere when we weren't fishing or hunting. I tried a sling but I was pathetic, I needed my RA Tuna gun. Our guide prepared conch for us after we furnished him the shells.

The 2nd year, we went during lobster season. They were hard to come by and small compared to what we used to get in the USVI. It was a bad trip for me, I had damaged my sacro stepping in a hole wade fishing. I was in pain the whole trip. Bob got a Nassau so we finally had a fish dinner.

CARDIOLOGY

A few months later Glenda and I were walking at Brazos Bend Park when we came to a hill. I started up the hill and ran out of breath. Glenda walked right past me. Wait a minute, something is wrong. I made an appointment with Glenda's cardiologist and had my heart checked. They ran a bunch of tests on fancy machines, did a sitting EKG and told me I wasn't getting enough oxygen. Then I had to go in to the catheter lab where they go in your vein and look inside your heart. I was told you have 3 main circuits in your heart, 2 of mine were blocked. (circuits, not vessels, that was strange to me) My blood vessels have little cholesterol.

We were leaving the building when the "electrician" called Glenda on her cell. Where are you? He asked, hold still, I'll be right there. He ran through the huge Methodist Hospital and found us. He was the pace-maker doc. (electrophysiologist) He sent me back upstairs to be wired up for observation. You have a transmitter on your hip and they watch your heart pulse. They have to watch you for a month to get insurance to pay for a pacemaker.

Three days later the doctor, (not a nurse) called in the morning and told Glenda my heart had stopped last night, can you bring him in. I wasn't scheduled so we went in on standby and waited several hours. You are awake during the procedure and it is freezing in the cath lab. I liked that doc and I was blabbing away, he finally told me to shut up for a minute, he had to concentrate. I have 2 leads in my vessels from the pacemaker to my heart. There are little hooks on the end which are pushed into the heart muscle, then lightly tugged back to assure attachment. He later told us my episode in the Bahamas was most likely caused by my heart. My body tried to vomit when I was perfectly calm, my heart violently reacted needing oxygen. You can't vomit underwater freediving without drowning.

WES TUNNEL

A well known Phd Marine Biologist at Corpus Christi A&M.passed away. His family was close to Glenda and Billy Causey when Billy was studying at the old UCC in the 60's. (University of Corpus Christi) There was going to be a large service and many of the original classmates would be there. Glenda wanted to attend. I loved my pacemaker. It made me feel young again. I located the former high school kid Jimmy Heidland and I used to take diving for free. And asked about a trip out of Port Aransas. Steve Shaw was all excited, he lived in Ingleside and wanted us to stay with him.

It was a lovely service. I saw Jim Copeland, Bob Turner and of course Glenda's ex Billy Causey. Billy had risen through the food chain and was head of all reef systems for NOAA.

Steve had a large catamaran with 2 large out boards. It was quite a boat and looked brand new and rarely used. He called his friend Mikey and away we went to the jetties. Steve and Mikey were scuba divers and had assorted scuba gear laid out, which they made no attempt to use. I thought it was strange he stopped at the jetties. There was a cold brown thermocline on the bottom and I could see little. Glenda asked Steve if he wouldn't mind running offshore, we had money for fuel. Grudgingly Steve wanderd out to some close rigs, still in brown water. We urged him offshore and finally he ran out about 16 miles to a lovely rig with good GOM vis. I call good green. I was anxious to try my pacemaker, I had been swimming pool laps with no problem. There was a school of gray snapper at the rig and we started playing the game. Steve never tied up, he just sat away from the rig and waited for me. I got a nice gray (mangrove) and came to the boat very happy. No ice? Where do you usually put your fish Steve? I got back in with Glenda, we were enjoying ourselves just being in mother ocean again. The fish had gotten spookier and harder to approach. I had my eye on one big one and we played the game for a while. Most of the school had moved outside the rig by now and gone deeper. I was getting embarrassed that Steve was just out here for me and waiting for us. I told Glenda I'm going outside the rig and dive straight down as deep as I can and try for a shot. My RA is deadly accurate, I have shot grays like this before. I did a long breath up and dove straight down towards the school as they went deeper, finally picking one out, pointing at a low profile (fish not sideways) and shot straight down. Got it.

Back at the boat I was very happy, I could dive again, my PM was a wonderful gift for my heart. Designed by an electrical engineer, of course. Steve and Mikey never attempted to dive. We had a blast and were ready for more but Steve headed in. Steve had a ss gun he had made in school that weighed about 15 pounds. Maybe good back in the jewfish days, just shoot and drop it and hang on to your rope. I asked him "what do you shoot with that heavy gun" he said "anything I want" Back at Steves house I asked him where to clean the fish, no clue. Where to put scraps? Just throw them on the ground, critters will get them at night. (stinking mess the next morning) I came to the conclusion that the boat and scuba gear were just for show and they didn't even dive any more. This was all just to impress us with all of his success. How sad.

Living on retirement now I have limited resources to pay for charters. I hope to travel to Alabama and dive with my son again in the old Seacraft the Pequod 3. It has a great boarding ladder that an old man can still climb, if he's not too cold.

Those 2 mango's are the last 2 snapper I have shot. Willie Powers says I still have one more snapper in me at 86.

FACEBOOK FRIENDS AND YOUTUBE VIDEOS

I do a lot of spearfishing now vicariously through others videos. For my 85th birthday Glenda and I chartered 2 days of redfishing in Venice La with Paul Miller. Paul is a great spearo. We caught fish until our arms wore out. I am FB friends with the whole Miller clan. Paul, Mark, David and even David's wife Caroline.

As I type a young spearo in Flour Bluff Tx, Braden Sherron had an epic March wahoo trip out of Port Aransas. The ocean was filled with hoo. Braden got his PB 80 lbs plus 6 more. There were so many they could miss and make mistakes and still get more shots and more hoo. Strange to me, they had no sharks to contend with.

David Ochoa makes the most beautiful spearfishing videos I have ever seen. I also like to follow and watch the Key West Waterman Aaron Young.

The screaming young stud he man videos are a turn off for me. Spearfishing is a dance with mother ocean, not an attack.

Old Freedive List friends I am still in touch with on a regular basis are: Daryl Wong, Brandi Easter, David Wittington, Reid Quinlan, Kurt Bickel, Troy Williams, Mike Wade, Bob Hurwitz, Tim Hatler, Peppe Di Mauro, Julie and Jill Riffe, Ray Ayus.

The passing of Carla Sue Hanson came as a shock to me. She was the secretary for AIDA and so sweet and kind to someone she never met. FB has brought many with common interests together. However, politics caused many to leave.

Black coral shark

Glenda the love of my life

EPILOGUE

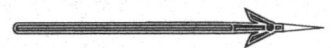

For anyone interested in my early life.

Born in San Diego Calif. 1936. Moved to Pearl City age 5 with the Pacific Fleet. Lived on the waterfront during the attack. (I have a statement I make every year on FB. PH was bait to enrage the pacifist public and save Europe) My mother moved to Chicago to be near her oldest brother. We were in a slum area and I ran with gangs until child welfare was going to take me. My mother sent me to live with my grandparents in Malad Idaho. I was in the 3rd grade. My grandmother's love and the Mormon way fixed a troubled kid. My mother died when I was 8. My father and step mother came and got me and my sister in 1946. We lived in Peary Place, Corpus Christi Navy housing. I was 10 years old. I graduated from W.B. Ray High School in1955. My sister, Mildred Krnavek died from diabetes in her 60's.

My father's mother married a dentist in St Louis, Frank Cates. They divorced and the Morrow family moved to Fort Worth. She then married Crawford Edwards (Edwards Ranch) and died shortly after in the 1919 Flu epidemic. My dad was then moved to the bunkhouse with the cowboys. He lied about his age and joined the Navy at age 16. He was a bisexual pedophile, and served 2 years in prison. One of my happiest days was leaving home.

My mothers parents (Buehler) left the coal mines of Park City Utah to homestead in Daniels Idaho, Near Malad. They suffered great hardship in the beginning.

I married Jackie Austin in 1957. Married Glenda Russell Cates in 1976.

Jackie, the mother of my 3 boys, died at age 64 with lung cancer. The whole family died from ailments caused by smoking. I despise cigarettes and what this drug does to the addicted.

Sadly or wonderfully, we all have our parents genes. I sincerely hope that you have wonderful parents. I was always jealous of kids with great parents.

I tried to tell my story with minimal personal drama and conflict. I had to put a little in, for context. Why and when we were here or there. I have no ill will towards anyone in this story.

Life is hard and we have to live with our choices. I do not believe in a magic sky god. I like all good people and understand the popularity of religion for community good. The basis of religion is forgiveness, and I humbly beg those I have harmed to forgive me.